Numerical Engineering Optimization

Andreas Öchsner · Resam Makvandi

Numerical Engineering Optimization

Application of the Computer Algebra System Maxima

 Springer

Andreas Öchsner
Faculty of Mechanical Engineering
Esslingen University of Applied Sciences
Esslingen am Neckar, Baden-Württemberg
Germany

Resam Makvandi
Institute of Mechanics
Otto von Guericke University Magdeburg
Magdeburg, Sachsen-Anhalt, Germany

ISBN 978-3-030-43390-1   ISBN 978-3-030-43388-8   (eBook)
https://doi.org/10.1007/978-3-030-43388-8

This Springer imprint is published by the registered company Springer Nature Switzerland AG
The registered company address is: Gewerbestrasse 11, 6330 Cham, Switzerland

Preface

This book is intended as a study aid for numerical optimization techniques in undergraduate as well as postgraduate degree courses in mechanical engineering at universities. Such procedures become more and more important in the context of lightweight design where a weight reduction may directly result, for example in the case of automotive or aerospace industry, in a lower fuel consumption and the corresponding reduction of operational costs as well as beneficial effects on the environment. Based on the free computer algebra system Maxima, we offer routines to numerically solve problems from the context of engineering mathematics as well as applications taken from the classical courses of strength of materials. The mechanical theories focus on the classical one-dimensional structural elements, i.e. springs, bars, and Euler–Bernoulli beams. Focusing on simple structural members reduces the complexity of the numerical framework, and the resulting design space is limited to a low number of variables. The use of a computer algebra system and the incorporated functions, e.g. for derivatives or equation solving, allows to focus more on the methodology of the optimization methods and not on the standard procedures. Some of the provided examples should be also solved in a graphical approach based on the objective function and the corresponding design space to better understand the computer implementation.

We look forward to receiving some comments and suggestions for the next edition of this textbook.

Esslingen am Neckar, Germany
Magdeburg, Germany
December 2019

Andreas Öchsner
Resam Makvandi

Contents

Chapter 1
Introduction

Abstract This chapter shortly introduces into the context of mathematical optimization problems. The basic mathematical notation is provided and the major idea of a numerical optimization problem is sketched. The second part summarizes a few basic operations of the computer algebra system Maxima, as well as the internet links to download the software. The few topics covered are basic arithmetics, definition of variables and functions etc. For a comprehensive introduction, the reader is referred to the available literature.

1.1 Optimization Problems

The general problem statement of a numerical optimization procedure [9] is to minimize the objective function F, e.g. the weight of a mechanical structure or the costs of a project,

$$F(X) = F(X_1, X_2, X_3, \dots, X_n),\tag{1.1}$$

subject to the following constraints

$$g_j(X) \le 0 \qquad\qquad j = 1, m \qquad \text{inequality constraints},\tag{1.2}$$
$$h_k(X) = 0 \qquad\qquad k = 1, l \qquad \text{equality constraints},\tag{1.3}$$
$$X_i^{\min} \le X_i \le X_i^{\max} \qquad i = 1, n \qquad \text{side constraints},\tag{1.4}$$

where the column matrix of design variables, e.g. geometrical dimensions of a mechanical structure, is given by:

$$X = \begin{Bmatrix} X_1 \\ X_2 \\ X_3 \\ \vdots \\ X_n \end{Bmatrix}.\tag{1.5}$$

A. Öchsner and R. Makvandi, *Numerical Engineering Optimization*, https://doi.org/10.1007/978-3-030-43388-8_1

In the case of a one-dimensional objective function, i.e. $\mathbf{X} \rightarrow X$, we end up with the following set of equations:

$$F(X) \qquad\qquad\qquad\qquad\qquad\qquad \text{objective function}, \qquad (1.6)$$

$$g_j(X) \leq 0 \qquad\qquad j = 1, m \qquad \text{inequality constraints}, \qquad (1.7)$$

$$h_k(X) = 0 \qquad\qquad k = 1, l \qquad \text{equality constraints}, \qquad (1.8)$$

$$X^{\min} \leq X \leq X^{\max} \qquad\qquad\qquad \text{side constraint}. \qquad (1.9)$$

To introduce different numerical schemes to find the minimum of the objective function, we first consider unconstrained one-dimensional objective functions (see Chap. 2), i.e. to find the minimum of the objective function $F(X)$ in certain boundaries, i.e. $[X^{\min} \leq X_{\text{extr}} \leq X^{\max}]$. The consideration of inequality ($g_i \leq 0$) and equality ($h_k = 0$) constraints is then added in Chap. 3. These one-dimensional strategies are then generalized to n-dimensional unconstrained problems (see Chap. 4) and then finally as constrained problems (see Chap. 5). With this book, we follow an educational approach, i.e. combining engineering foundation with Maxima, which we first introduced in the scope of the finite element method, see [7, 8].

1.2 Maxima—A Computer Algebra System

The computer algebra system Maxima was originally developed in the late 1960s and earlier 1970s at MIT under the name 'Macsyma'. The historical development is excellently summarized in the article by Moses [4]. Compared to many commercial alternatives such as Maple, Matlab or Mathematica, Maxima is distributed under the GNU General Public License (GPL) and thus a free software. In addition, it is recommended to use in the following the wxMaxima graphical user interface (GUI), which is also distributed under the GNU GPL. The routines used in the following chapters are optimized for this GUI. Both programs can be downloaded from the links in Table 1.1.

A few basic operations in Maxima are explained in the following without any attempt for completeness. The interested reader may find elsewhere further links and examples in order to study the functionality of Maxima [1–3]. Particularly worth mentioning is the web page 'Maxima by Example' by Woollett [10]. Let us now start with the basic arithmetic operations as outlined in the following Listing 1.1.

Table 1.1 Links to download the installation packages

Program	Link
Maxima	http://maxima.sourceforge.net/download.html
wxMaxima	http://andrejv.github.io/wxmaxima/

Basic arithmetics: $+, -, *, /$

```
(% i1)    1 + 2;
(% o1)    3


(% i2)    1 - 2;
(% o2)    -1


(% i3)    1 * 2;
(% o3)    2


(% i4)    1 / 2;
(% o4)    ½
```

Module 1.1: Basic arithmetics

We can see from the above example that all expressions entered into Maxima must end with a semicolon ';'. Alternatively, the '$\$$' character can be used at the end of a statement to suppress the output of that line. Some pre-defined functions in Maxima are collected in Table 1.2. It should be noted here that the arguments of the trigonometric functions must be given in radians.

Another characteristic is that Maxima tries to provide symbolic results, which include fractions, square roots etc. The function float(...) can be used to obtain a floating-point representation as shown in the following listing. The most recent result can be recalled with the percentage operator (%).

The value of a variable is entered by the use of a colon ':', see the following listing. To clear a value from an assigned variable, the command kill(...), or for all variables kill(all), is used.

Some predefined constants in Maxima are '%e' (i.e., the base of the natural logarithm; e = 2.718281...), '%pi' (i.e., the ratio of the perimeter of a circle to its diameter; π = 3.141592...), and '%i' (i.e., the imaginary unit; $\sqrt{-1}$).

Table 1.2 Some pre-defined functions in Maxima

Command	Meaning	Command	Meaning
sqrt(...)	Square root	sin(...)	Sine
exp(...)	Exponential	cos(...)	Cosine
log(...)	Natural logarithm	tan(...)	Tangent
abs(...)	Absolute value	cot(...)	Cotangent

```
(% i1)      1/2;
(% o1)      ½

(% i2)      float(%);
(% o2)      0.5

(% i3)      float(1/2);
(% o3)      0.5
```

Module 1.2: Use of the function float(...)

```
Definition of variables and calculations:
(% i2)      a: 3;
            b: 4;

(a)         3
(b)         4

(% i3)      c: a + b;

(c)         7
```

Module 1.3: Definition of variables

A function is defined by the use of a colon followed by the equal sign ':=', see the following Listing 1.4. In regard to the naming of variables and functions, it should be noted that the names must start with a letter and may contain numbers or even an underscore. The following way of defining a function uses the general structure

$$f(x) := (expr1, expr2, ..., exprn),$$

where the value of exprn is returned by the function $f(x)$.

Purpose: Calculates the sum of two numbers.
Input(s): Values a and b.
Output: Sum of a and b stored in variable c

```
(% i1)    kill(all)$
(% i1)    summation(a,b) := a+b $
(% i3)    a:3 $
          b:5 $
(% i4)    c : summation(a,b);

(c)       8
```

Module 1.4: Definition of a function

The following example (see Module 1.5) shows the definition of a function under consideration of a block structure. The block structure allows to make a return from some expression inside the function, to return multiple values, and to introduce local variables.

Purpose: Calculates the length of a straight line between two points.
Input(s): Coordinates of the end points.
Output(s): Length of the line connecting the points.

```
⟶    LineLength(ncoor):=
     block([x1,x2,y1,y2,x21,y21,L,LL],
     [[x1,y1],[x2,y2]] : ncoor,
     [x21,y21] : [x2-x1,y2-y1],
     LL : (x21^2+y21^2),
     L : sqrt(LL),
     return(L)
     )$


⟶    LineLength([[0,0],[1,1]]);

(% o2)    √2
```

Module 1.5: Definition of a function under consideration of a block structure

Let us mention next the command **ratsimp(...)**, which allows to simplify algebraic expressions.[1] The following Listing 1.6 illustrates the simplification of the expressions

$$f_1(x) = 2x^2 + (x - 1)^2,$$ (1.10)

$$f_2(x) = \frac{x^2 + 4x - 3x^2 + 1}{(x - 1)^2 - 1 + 2x}.$$ (1.11)

Manipulation of algebraic expressions:

(% i2)   f1: 2*x^2+(x-1)^2$
          f2: (x^2+4*x-3*x^2+1)/((x-1)^2-1+2*x)$

(% i3)   ratsimp(f1);

(% o3) $3x^2 - 2x + 1$

(% i4)   ratsimp(f2);

(% o4) $-\frac{2x^2 - 4x - 1}{x^2}$

Module 1.6: Manipulation and simplification of algebraic expressions

Matrix operations and the corresponding solution of linear system of equations in matrix form are of utmost importance in the context of engineering mathematics. The following Listing 1.7 illustrates the solution procedure of a linear system of equations [6]:

$$\begin{bmatrix} \dfrac{15AE}{L} & -\dfrac{6AE}{L} & 0 \\ -\dfrac{6AE}{L} & \dfrac{9AE}{L} & -\dfrac{3AE}{L} \\ 0 & -\dfrac{3AE}{L} & \dfrac{3AE}{L} \end{bmatrix} \begin{bmatrix} x_1 \\ x_2 \\ x_3 \end{bmatrix} = \begin{bmatrix} 0 \\ 0 \\ F_0 \end{bmatrix}.$$ (1.12)

---

[1] Further commands for the manipulation of algebraic expressions can be found in the references given on the web page [3].

Load the library mbe5.mac which contains routines for matrix operations (optional, depending on installation)

(% i1)
```
/*load("mbe5.mac")$*/
```

```
display2d:true$
```

Definition of the coefficient matrix Ac

(% i4) `Ac : 3*E*A/L*matrix([5,-2,0],[-2,3,-1],[0,-1,1])$`

```
print(" ")$
print(" ","Ac =", Ac)$
```

$$Ac = \begin{bmatrix} \dfrac{15AE}{L} & -\dfrac{6AE}{L} & 0 \\ -\dfrac{6AE}{L} & \dfrac{9AE}{L} & -\dfrac{3AE}{L} \\ 0 & -\dfrac{3AE}{L} & \dfrac{3AE}{L} \end{bmatrix}$$

Calculation of the inverse A_inv of the coefficient matrix Ac

(% i7) `A_inv : invert(Ac)$`

```
print(" ")$
print(" ","A_inv =", A_inv)$
```

$$A\_inv = \begin{bmatrix} \dfrac{L}{9AE} & \dfrac{L}{9AE} & \dfrac{L}{9AE} \\ \dfrac{L}{9AE} & \dfrac{5L}{18AE} & \dfrac{5L}{18AE} \\ \dfrac{L}{9AE} & \dfrac{5L}{18AE} & \dfrac{11L}{18AE} \end{bmatrix}$$

Definition of the right-hand side, i.e. load vector (column matrix)

(% i10) `rhs : matrix([0],[0],[F_0])$`

```
print(" ")$
print(" ","rhs =", rhs)$
```

$$rhs = \begin{bmatrix} 0 \\ 0 \\ F_0 \end{bmatrix}$$

Definition of the right-hand side, i.e. load vector (column matrix)

(% i10)   rhs : matrix([0],[0],[F_0])\$

          print(" ")\$
          print(" ","rhs =", rhs)\$

$$rhs = \begin{bmatrix} 0 \\ 0 \\ F_0 \end{bmatrix}$$

Solution of system, i.e. multiplication of the inverse with the right-hand side

(% i13)   x_sol : A_inv . rhs\$

          print(" ")\$
          print(" ","x_sol =", x_sol)\$

$$x\_sol = \begin{bmatrix} \dfrac{F_0 L}{9AE} \\[2mm] \dfrac{5F_0 L}{18AE} \\[2mm] \dfrac{11F_0 L}{18AE} \end{bmatrix}$$

Access single elements of the solution vector

(% i16)   print(" ")\$
          print("Elements of the solution vector:")\$
          for i:1 thru length(x_sol) do(
              print(" "),
              print(" ", concat('x_sol_,i), "=",
          args(x_sol)[i])
)\$

Elements of the solution vector:

$$x\_sol_1 = [\frac{F_0 L}{9AE}]$$

$$x\_sol_2 = [\frac{5F_0 L}{18AE}]$$

$$x\_sol_3 = [\frac{11F_0 L}{18AE}]$$

Module 1.7: Solution of a linear system of equations in matrix form

The following example (see Module 1.8) shows the symbolic integration of integrals. The task is to calculate the integrals

$$\int_0^{\frac{L}{2}} \frac{(M_1(x))^2}{2EI_1(x)} \, dx \, , \tag{1.13}$$

$$\int_0^{\frac{L}{2}} \frac{16}{25} \varrho \, (d_1(x))^2 \, dx \, , \tag{1.14}$$

where the functions are given as follows:

$$M_1(x) = \frac{3F_0L}{4} \left(1 - \frac{2x}{3L} \right) , \tag{1.15}$$

$$d_1(x) = d_0 \left(1 - \frac{2x}{3L} \right)^{\frac{1}{3}} , \tag{1.16}$$

$$I_1(x) = \frac{136}{1875} \, (d_1(x))^4 \, . \tag{1.17}$$

The following example (see Module 1.9) shows the symbolic derivation of functions. The task is to calculate the first- and second-order derivative of the function

$$f(x) = 0.05 \times (3 - x)^4 + 1 . \tag{1.18}$$

In the context of optimization procedures, the calculation of the gradient operator (see Eq. 4.1) and the Hessian matrix (see Eq. 4.2) is required for first- and second-order methods, respectively. The following example (see Module 1.10) shows the symbolic derivation of the gradient operator and the Hessian matrix for the function

$$f(x, y, z) = x^3 + \frac{1}{x} + y + \frac{1}{y} - z^4 . \tag{1.19}$$

At the end of this section, the functionality to generally solve differential equations is shown. This allows to derive the general solutions for rods, Euler-Bernoulli beams or Timoshenko beams, see [5]. Thus, numerical solutions based on the finite element approach can be compared to the analytical, i.e. exact, solution for simple problems. The different differential equations are given in the following Modules 1.11–1.13 for constant material and geometrical properties. We obtain for rods

$$EA \frac{d^2 u_x(x)}{dx^2} = -p_x , \tag{1.20}$$

Definitions:

($\%$i4) assume(L>0)\$
M_1(x) := 3*F_0*L/4*(1-2*x/(3*L))\$
d_1(x) := d_0*(1-2*x/(3*L))^(1/3)\$
I_1(x):=136/1875*d_1(x)^4\$

Evaluation of integrals:

($\%$i16) simp : false\$ /* disables automatic simplifications */
fpprintprec : 6\$ /* float numbers precision (only for printing) */

P_1 : 'integrate(M_1(x)^2/(2*E*I_1(x)),x,0,L/2)\$

P_2 : ev(integrate(M_1(x)^2/(2*E*I_1(x)),x,0,L/2),simp)\$
/* ev(ex,simp)Ã,Â :Ã,Â selective simplification */
P_3 : ev(float(P_2),simp)\$

M_1 : 'integrate(rho*16/25*d_1(x)^2,x,0,L/2)\$

M_2 : ev(integrate(rho*16/25*d_1(x)^2,x,0,L/2),simp)\$

M_3 : ev(float(M_2),simp)\$

print(" ")\$
print(P_1,"=",P_2,"=",P_3)\$
print(" ")\$
print(M_1,"=",M_2,"=",M_3)\$

$$\int_0^{\frac{L}{2}} \frac{\left(\frac{3F_0 L}{4}\left(1-\frac{2x}{3L}\right)\right)^2}{2E\left(\frac{136}{1875}\left(d_0\left(1-\frac{2x}{3L}\right)^{\frac{1}{3}}\right)^4\right)} \, dx = \frac{16875 F_0{}^2 L^2 \left(\frac{9L}{10} - \frac{2^{\frac{2}{3}} 3^{\frac{1}{3}} L}{5}\right)}{4352 E \, d_0{}^4} = \frac{1.71431 F_0{}^2 L^3}{E \, d_0{}^4}$$

$$\int_0^{\frac{L}{2}} \frac{\varrho 16}{25}\left(d_0\left(1-\frac{2x}{3L}\right)^{\frac{1}{3}}\right)^2 \, dx = \frac{16\left(\frac{9L}{10} - \frac{2^{\frac{2}{3}} 3^{\frac{1}{3}} L}{5}\right) d_0{}^2 \varrho}{25} = 0.282953 L \, d_0{}^2 \varrho$$

Module 1.8: Symbolic integration of integrals

Definitions:

(%i2)   fpprintprec : 6$ /* float numbers precision (only for printing) */
          f(x) := 0.05*(3-x)^4+1$

Evaluation of derivatives:

(%i9)   symb_df : 'diff(f(x),x,1)$
          symb_ddf : 'diff(f(x),x,2)$

          df : diff(f(x),x,1)$
          ddf : diff(f(x),x,2)$

          print(" ")$
          print(symb_df,"=",df)$
          print(symb_ddf,"=",ddf)$

$$\frac{\mathrm{d}}{\mathrm{d}x}\left(0.05(3-x)^4+1\right)=-0.2(3-x)^3$$

$$\frac{\mathrm{d}^2}{\mathrm{d}x^2}\left(0.05(3-x)^4+1\right)=0.6(3-x)^2$$

Module 1.9: Symbolic derivations of functions

Definitions:

(%i2)   load("my_funs.mac")$
          f(x,y,z) := x^3 + (1/x) + y + (1/y) - z^4$

Evaluation of Gradient and Hessian:

(%i10)   grad_f : gradient(f(x,y,z), [x,y,z])$
          hess_f : hessian(f(x,y,z), [x,y,z])$

          print(" ")$
          print(" Gradient of f(x,y,z):")$
          print(" ", grad_f)$
          print(" ")$
          print(" Hessian of f(x,y,z):")$
          print(" ", hess_f)$

Gradient of f(x,y,z):

$$[3x^2 - \frac{1}{x^2}, 1 - \frac{1}{y^2}, -4z^3]$$

Hessian of f(x,y,z):

$$\begin{pmatrix} 6x + \dfrac{2}{x^3} & 0 & 0 \\ 0 & \dfrac{2}{y^3} & 0 \\ 0 & 0 & -12z^2 \end{pmatrix}$$

Module 1.10: Symbolic derivations of the gradient operator and the Hessian matrix for the function (1.19)

Solution of DE for rod with constant tensile stiffness (EA) and constant distributed load p; command odeL requires to load the package odes.mac:

(%i9)   load("odes.mac")$

```
bar : EA*diff(u,x,2) = -p$
gen_sol : odeL(bar,u,x)$

print(" ")$
print("The differential equation:")$
print(" ", bar)$
print(" ")$
print("The general solution:")$
print(" ", gen_sol)$
```

The differential equation:

$$EA\left(\frac{d^2}{dx^2}u\right) = -p$$

The general solution:

$$u = -\frac{p x^2}{2EA} + C2x + C1$$

Module 1.11: General solution of differential equations: rod element, see Eq. (1.20)

Solution of DE for EB beam with constant bending stiffness (EI) and constant distributed load q; command odeL requires to load the package odes.mac:

(%i9)   load("odes.mac")$

```
Ebeam : EI*diff(u,x,4) = q$
gen_sol : odeL(Ebeam,u,x)$

print(" ")$
print("The differential equation:")$
print(" ", Ebeam)$
print(" ")$
print("The general solution:")$
print(" ", gen_sol)$
```

The differential equation:

$$EI\left(\frac{d^4}{dx^4}u\right) = q$$

The general solution:

$$u = \frac{q x^4}{24EI} + C4x^3 + C3x^2 + C2x + C1$$

Module 1.12: General solution of differential equations: Euler-Bernoulli beam element, see Eq. (1.21)

Solution of the coupled DE for Timoshenko beam for kAG, EI, q, m = const.:

(%i14)   eqn_1: -kAG*'diff(u(x),x,2) = q+kAG*'diff(phi(x),x)$
         eqn_2: -kAG*'diff(u(x),x) = -m-EI*'diff(phi(x),x,2)+kAG*phi(x)$

         sol : desolve([eqn_1, eqn_2], [u(x),phi(x)])$

         print(" ")$
         print("Equations:")$
         print(" ", eqn_1)$
         print(" ")$
         print(" ", eqn_2)$
         print(" ")$
         print(" ")$
         print("Solutions:")$
         print(ratsimp(sol[1]))$
         print(" ")$
         print(ratsimp(sol[2]))$

Equations:

$$-kAG\left(\frac{\mathrm{d}^2}{\mathrm{d}x^2}\,\mathrm{u}(x)\right) = kAG\left(\frac{\mathrm{d}}{\mathrm{d}x}\,phi(x)\right) + q$$

$$-kAG\left(\frac{\mathrm{d}}{\mathrm{d}x}\,\mathrm{u}(x)\right) = -EI\left(\frac{\mathrm{d}^2}{\mathrm{d}x^2}\phi(x)\right) + kAG\,\phi(x) - m$$

Solutions:

$$\mathrm{u}(x) = -\frac{\begin{array}{c}\left(4kAG^2\,x^3 - 24EI\,kAGx\right)\left(\frac{\mathrm{d}}{\mathrm{d}x}\,\mathrm{u}(x)\big|_{x=0}\right) + 12EI\,kAG\,x^2\left(\frac{\mathrm{d}}{\mathrm{d}x}\phi(x)\big|_{x=0}\right) \\ -kAGq\,x^4 + \left(4\phi(0)\,kAG^2 - 4kAGm\right)x^3 + 12EIq\,x^2 - 24\,\mathrm{u}(0)\,EI\,kAG\end{array}}{24EI\,kAG}$$

$$\phi(x) = \frac{3kAG\,x^2\left(\frac{\mathrm{d}}{\mathrm{d}x}\,\mathrm{u}(x)\big|_{x=0}\right) + 6EIx\left(\frac{\mathrm{d}}{\mathrm{d}x}\phi(x)\big|_{x=0}\right) - q\,x^3 + (3\phi(0)\,kAG - 3m)\,x^2 + 6\phi(0)\,EI}{6EI}$$

Module 1.13: General solution of differential equations: Timoshenko beam element, see Eqs. (1.22) to (1.23) and for Euler-Bernoulli beams

$$EI_y \frac{\mathrm{d}^4 u_z(x)}{\mathrm{d}x^4} = q_z \,,$$

(1.21)

and the coupled system of equations for Timoshenko beams:

$$EI_y \frac{\mathrm{d}^2 \phi_y}{\mathrm{d}x^2} - k_s G A \left(\frac{\mathrm{d}u_z}{\mathrm{d}x} + \phi_y \right) + m_z = 0$$

(1.22)

$$-k_s G A \left(\frac{\mathrm{d}^2 u_z}{\mathrm{d}x^2} + \frac{\mathrm{d}\phi_y}{\mathrm{d}x} \right) - q_z = 0 \,.$$

(1.23)

References

1. Hannan Z (2019) wxMaxima for calculus I. https://wxmaximafor.wordpress.com. Cited 15 Jan 2019
2. Hannan Z (2019) wxMaxima for calculus II. https://wxmaximafor.wordpress.com. Cited 15 Jan 2019
3. Maxima Dokumentation (2018) http://maxima.sourceforge.net/documentation.html. Cited 23 March 2018
4. Moses J (2012) Macsyma: A personal history. J Symb Comput 47:123–130
5. Öchsner A (2014) Elasto-plasticity of frame structure elements: modelling and simulation of rods and beams. Springer, Berlin
6. Öchsner A (2016) Computational statics and dynamics: an introduction based on the finite element method. Springer, Singapore
7. Öchsner A, Makvandi R (2019) Finite elements for truss and frame structures: an introduction based on the computer algebra system Maxima. Springer, Cham
8. Öchsner A, Makvandi R (2019) Finite elements using Maxima: theory and routines for rods and beams. Springer, Cham
9. Vanderplaats GN (1999) Numerical optimization techniques for engineering design. Vanderplaats Research and Development, Colorado Springs
10. Woollett EL (2018) Maxima by example. http://web.csulb.edu/~woollett/. Cited 23 March 2018

Chapter 2
Unconstrained Functions of One Variable

Abstract This chapter introduces three classical numerical methods to find the minimum of a unimodal function of one variable. The first two methods, i.e., the golden section algorithm and the brute-force algorithm, are typical representatives of zero-order methods which only require functional evaluations of the objective function. The third method, i.e., Newton's method, is a typical representative of second-order methods and requires the evaluation of the first-and second-order derivatives of the objective function.

2.1 Golden Section Algorithm

The main idea of the golden section algorithm is to reduce an initial interval $[X_{\min}, X_{\max}]$ to a predefined small size which contains the minimum, see [2] for details. The actual approach is now that the reduction between two subsequent intervals is kept equal to the golden ratio $\tau = \frac{3-\sqrt{5}}{2} \approx 0.381966$:

$$\tau = \frac{\text{reduction in interval size}}{\text{previous interval size}}. \tag{2.1}$$

From this relationship, it can be concluded that the ratio between the actual and previous interval sizes can be expressed as follows:

$$1 - \tau = \frac{\text{actual interval size}}{\text{previous interval size}}. \tag{2.2}$$

The method solely requires functional evaluations, i.e. there is no need for any form of derivative to be calculated or approximated, nor do we need to inspect if the derivative is continuous or not. The objective function is evaluated first at its initial boundaries ($X_{\min}$ and $X_{\max}$) and then the interval is reduced and two inner points (X_1 and X_2) are determined based on the golden ratio. The objective function is also

A. Öchsner and R. Makvandi, *Numerical Engineering Optimization*, https://doi.org/10.1007/978-3-030-43388-8_2

evaluated at these inner points and based on the assumption of a unimodal function,[1] a smaller interval is obtained. The abscissa which corresponds to the larger value of $F(X_1)$ and $F(X_2)$ forms the new boundary and the algorithm is continued in the same manner, see Fig. 2.1.

Let us use in the following the relative tolerance ε to specify the convergence of the algorithm:

$$\varepsilon = \frac{\text{actual interval size}}{\text{initial interval size}} = \frac{\Delta X}{X_{\max} - X_{\min}}, \tag{2.3}$$

where ΔX can be also referred to as the absolute tolerance. To reach an interval with a relative tolerance $\leq \varepsilon$, we need to fulfill the relation

$$(1 - \tau)^{N-3} \leq \varepsilon, \tag{2.4}$$

where N is the total number of function evaluations (F), including the initial three. Solving for N, we finally obtain the following relationship, which can be used to formulate the convergence criterion:

$$N(\varepsilon) = \frac{\ln(\varepsilon)}{\ln(1 - \tau)} + 3. \tag{2.5}$$

It should be noted here that the absolute number of iterations of the algorithm shown in Fig. 2.1 is equal to $N - 3$.

Furthermore, it should be noted here that the difference of the two supporting points, i.e. X_1 and X_2, gives the following ratio with respect to the previous interval size:

$$\frac{X_2 - X_1}{X_{\max} - X_{\min}} = 1 - 2\tau. \tag{2.6}$$

2.1 Numerical Determination of a Minimum for an Unconstrained Function

Determine based on the golden section method the minimum of the function

$$F(X) = \tfrac{1}{2}(X - 2)^2 + 1 \tag{2.7}$$

in the range $0 \leq X \leq 5$. Assign for the relative tolerance the value $\varepsilon = 0.005$.

2.1 Solution

The following Listing 2.1 shows the entire wxMaxima code for the determination of the minimum of function (2.7). The change of the intervals is illustrated in Fig. 2.2 for the first two iterations.

---

[1]In this context, a unimodal function is a function which has in the considered interval only a single minimum.

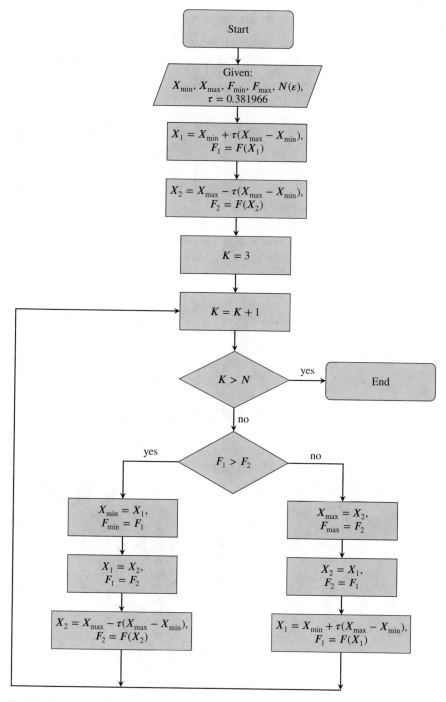

Fig. 2.1 Flowchart of the golden section algorithm for an unconstrained minimum, adapted from [2]

Fig. 2.2 Golden section
method for the example
$F(X) = \frac{1}{2}(X - 2)^2 + 1$.
Exact solution for the
minimum: $X_{\text{extr}} = 2.0$

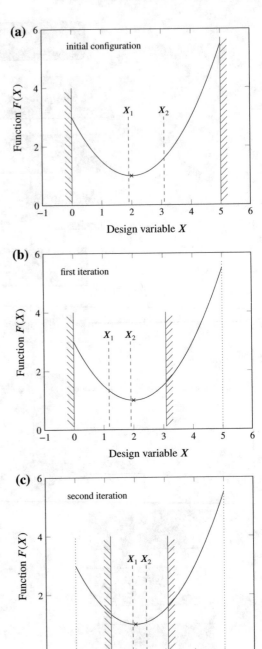

```
(% i2)    load("my_funs.mac")$
          load("engineering-format")$


(% i9)    func(x) := 0.5*(x-2)^2+1 $
          xmin : 0$
          xmax : 5$
          eps : 0.005$

          N : round((log(eps)/(log(1-0.381966))+3))$

          print("N =", N)$
          gss(xmin, xmax, N)$

N = 14
```

K	x_min	x_1	x_2	x_max	f_min	f_1	f_2	f_max
3	0.0000E–0	1.9098E+0	3.0902E+0	5.0000E+0	3.0000E+0	1.0041E+0	1.5942E+0	5.5000E+0
4	0.0000E–0	1.1803E+0	1.9098E+0	3.0902E+0	3.0000E+0	1.3359E+0	1.0041E+0	1.5942E+0
5	1.1803E+0	1.9098E+0	2.3607E+0	3.0902E+0	1.3359E+0	1.0041E+0	1.0650E+0	1.5942E+0
6	1.1803E+0	1.6312E+0	1.9098E+0	2.3607E+0	1.3359E+0	1.0680E+0	1.0041E+0	1.0650E+0
7	1.6312E+0	1.9098E+0	2.0820E+0	2.3607E+0	1.0680E+0	1.0041E+0	1.0034E+0	1.0650E+0
8	1.9098E+0	2.0820E+0	2.1885E+0	2.3607E+0	1.0041E+0	1.0034E+0	1.0178E+0	1.0650E+0
9	1.9098E+0	2.0163E+0	2.0820E+0	2.1885E+0	1.0041E+0	1.0001E+0	1.0034E+0	1.0178E+0
10	1.9098E+0	1.9756E+0	2.0163E+0	2.0820E+0	1.0041E+0	1.0003E+0	1.0001E+0	1.0034E+0
11	1.9756E+0	2.0163E+0	2.0414E+0	2.0820E+0	1.0003E+0	1.0001E+0	1.0009E+0	1.0034E+0
12	1.9756E+0	2.0007E+0	2.0163E+0	2.0414E+0	1.0003E+0	1.0000E+0	1.0001E+0	1.0009E+0
13	1.9756E+0	1.9911E+0	2.0007E+0	2.0163E+0	1.0003E+0	1.0000E+0	1.0000E+0	1.0001E+0
14	1.9911E+0	2.0007E+0	2.0067E+0	2.0163E+0	1.0000E+0	1.0000E+0	1.0000E+0	1.0001E+0

Module 2.1: Numerical determination of the minimum for the function $F(X) = \frac{1}{2}(X - 2)^2 + 1$ in the range $0 \leq X \leq 5$ based on the golden section method. Exact solution for the minimum: $X_{extr} = 2.0$.

2.2 Brute-Force or Exhaustive Search Algorithm

Let us assume again a unimodal function, i.e. a function that has only one minimum in the considered interval, see Fig. 2.3.

A simple numerical search algorithm for the minimum can be constructed in the following manner: Start from the left-hand boundary (X_{min}) and calculate in *small* constant steps (Δh) the functional values of the objective function (F). As soon as we fulfill for three consecutive points 0, 1 and 2 the condition

$$F(X_0) \geq F(X_1) \leq F(X_2), \tag{2.8}$$

Fig. 2.3 Initial configuration
for a brute-force approach
(version 1)

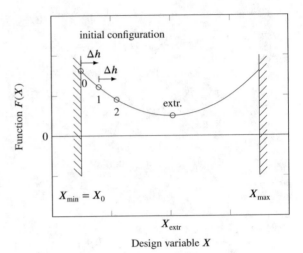

the minimum is localized within the interval $[X_0, X_2]$. This procedure is known in
literature as brute-force or exhaustive search. The flowchart in Fig. 2.4 indicates the
corresponding algorithm.

An alternative and sometimes faster modification of the algorithm in Fig. 2.4 can
be obtained by locating the start value (X_0) somewhere within the interval.[2] The next
task is then to decide if the search should move from this start point to the left-or
right-hand side of the interval. Naturally, the search will continue in the direction of
the descending function, see Fig. 2.6 for details.

In order to reduce the required iterations, approaches with variable step sizes can
be introduced, see Fig. 2.7. The variable step size, which is expressed in general by
$\Delta h^{(i)}$, can be written as

$$\Delta h^{(i)} = \alpha^{(i)} \times \Delta h^{(i-1)} \quad \text{for} \quad i > 0. \tag{2.9}$$

where $\alpha^{(i)}$ is a scaling parameter. In case that we set $\alpha^{(i)} = 1$, the classical brute
force algorithm with constant step size is recovered, see Fig. 2.6. In case that we set
$\alpha^{(i)} < 0$, the interval size reduces during each iteration, while $\alpha^{(i)} > 0$ results in a
larger interval for each iteration step. Alternatively, the interval size can be increased
based on the Fibonacci sequence,[3] i.e. $\alpha^{(i)} = 1, 1, 2, 3, 5, 8, 13, \ldots (i > 0)$.

---

[2]It is obvious that the choice of the start value has a significant influence on the required steps to
locate the minimum. Different initial configurations are illustrated in Fig. 2.5.

[3]The rule for the generation of the Fibonacci numbers is $F_n = F_{n-1} + F_{n-2}$ with $F_0 = 0$ and
$F_1 = 1$.

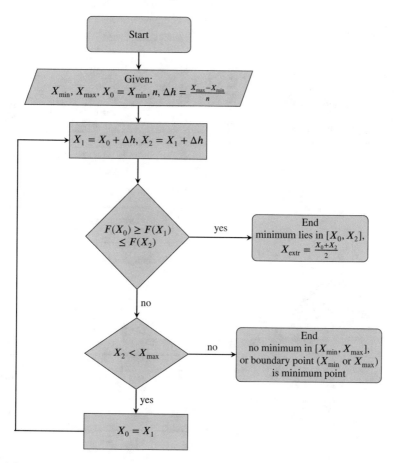

Fig. 2.4 Flowchart of the brute-force algorithm (version 1) for an unconstrained minimum

2.2 Numerical Determination of a Minimum for an Unconstrained Function

Determine based on the brute-force approach the minimum of the function

$$F(X) = \tfrac{1}{2}(X - 2)^2 + 1 \tag{2.10}$$

in the range $0 \leq X \leq 5$ (see Fig. 2.2 for a graphical representation of the function). Assign to version 2 different start values ($X_0 = 1.0, 1.5, 3.5$) and compare the results with the results of version 1. The step size should be increased in certain steps. Plot in addition for version 2 a diagram which shows the convergence rate, i.e. coordinate of the minimum X_{extr} and iteration number as a function of the step size parameter n, for the case $X_0 = 1.0$.

Fig. 2.5 Different initial
configurations of an
unconstrained function $F(X)$
in the scope of a brute-force
approach (version 2): **a** local
minimum within
$]X_{min}, X_{max}[$; **b** minimum at
left-hand boundary
$(X_{extr} = X_{min})$; **c** minimum
at right-hand boundary
$(X_{extr} = X_{max})$

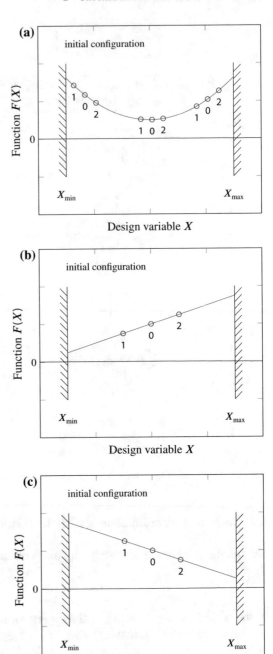

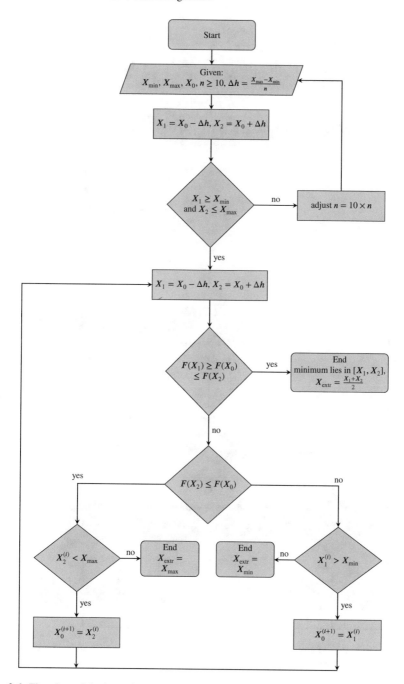

Fig. 2.6 Flowchart of the brute-force algorithm (version 2) for an unconstrained minimum

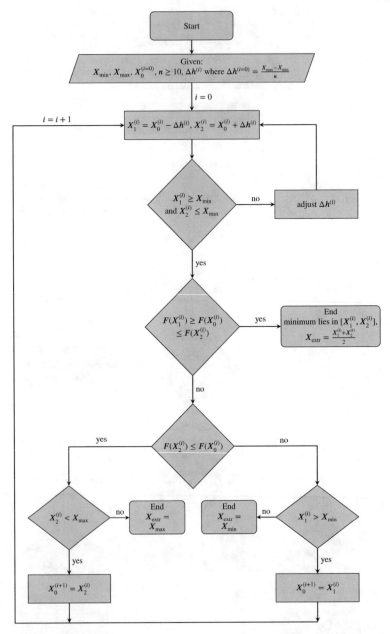

Fig. 2.7 Flowchart of the brute-force algorithm (version 3, i.e. with variable step size) for an unconstrained minimum

2.2 **Solution**

The following Listing 2.2 shows the entire wxMaxima code for the determination of
the minimum based on brute-force version 1 for the particular case of $n = 10$.

```
(% i2)    load("my_funs.mac")$
          load("engineering-format")$


(% i7)    func(x) := (1/2)*(x-2)^2+1 $
          xmin : 0$
          xmax : 5$

          n : 10$

          bf_ver1(xmin, xmax, n)$

          minimum lies in [ 1.5000e+0, 2.5000e+0]
          X_extr = 2.0000e+0 ( i = 4 )
```

Module 2.2: Numerical determination of the minimum for the function $F(X) = \frac{1}{2}(X - 2)^2 + 1$ in the range $0 \le X \le 5$ based on the brute-force approach (version 1) for $n = 10$. Exact solution for the minimum: $X_{extr} = 2.0$.

Other values for a variation of parameter n are summarized in Table 2.1.
    The wxMaxima code for the application of brute-force method 2 is illustrated in
Listing 2.3.

Table 2.1 Summary of detected minimum values (exact value: 2.0) for different parameters n, i.e.
different step sizes (brute-force version 1)

n	X_{min}	X_{max}	X_{extr}	i
4	1.250	3.750	2.500	2
8	1.250	2.500	1.875	3
10	1.500	2.500	2.000	4
15	1.667	2.333	2.000	6

```
(% i2)    load("my_funs.mac")$
          load("engineering-format")$


(% i7)    func(x) := (1/2)*(x-2)^2+1 $
          xmin : 0$
          xmax : 5$
          x_0 : 1.5$

          n : 10$

          bf_ver2(xmin, xmax, x_0, n)$

          minimum lies in [ 2.0000e+0, 2.5000e+0]
          X_extr = 2.2500e+0 ( i = 2 )
```

Module 2.3: Numerical determination of the minimum for the function $F(X) = \frac{1}{2}(X-2)^2 + 1$ in the range $0 \leq X \leq 5$ based on the brute-force approach (version 2) for $n = 10$ and $X_0 = 1.5$. Exact solution for the minimum: $X_{extr} = 2.0$.

Other values for a variation of parameter n and the start value X_0 are summarized in Table 2.2. It can be seen that these parameters have a considerable influence on the detected minimum and that the parameter n should be sufficiently large to guarantee an acceptable result for the minimum.

The convergence rate of the algorithm is illustrated in Fig. 2.8, which shows the coordinate of the minimum and the required iteration steps as a function of the step size parameter.

2.3 Numerical Determination of the Minimum for Two Unconstrained Functions

Determine based on the brute-force approach the minimum of the linear functions

$$F(X) = X - 0.5 \tag{2.11}$$

and

$$F(X) = -X + 2.5 \tag{2.12}$$

in the range $0.75 \leq X \leq 2.25$, see Fig. 2.9. Assign to algorithm version 2 different start values (X_0) and compare the results with the results of version 1. The step size can be calculted based on $n = 10$.

Table 2.2 Summary of detected minimum values (exact value: 2.0) for different parameters n, i.e. different step size, and different start values X_0 (brute-force version 2)

n	X_{min}	X_{max}	X_{extr}	i
$X_0 = 1.0$				
10	2.000	2.500	2.250	3
15	2.000	2.333	2.167	4
20	2.000	2.250	2.125	5
25	2.000	2.200	2.100	6
30	2.000	2.167	2.083	7
35	2.000	2.143	2.071	8
40	2.000	2.125	2.063	9
$X_0 = 1.5$				
10	2.000	2.500	2.500	2
15	1.833	2.167	2.000	2
20	2.000	2.250	2.125	3
25	1.900	2.100	2.000	3
$X_0 = 3.5$				
10	2.000	2.500	2.250	4
15	2.167	2.500	2.333	5
20	2.000	2.250	2.125	7
25	2.100	2.300	2.200	8
30	2.000	2.167	2.083	10
35	2.929	2.071	2.000	12

2.3 Solution

The following Listing 2.4 shows the entire wxMaxima code for the determination of the minimum based on brute-force version 1. Based on the output, it can be concluded that the left-hand boundary, i.e. $X = 0.75$, is the minimum point of function (2.11) under the assumption that we have a unimodal function.

The following Listing 2.5 shows the entire wxMaxima code for the determination of the minimum based on brute-force version 2 for the function given in Eq. (2.11).

In a similar way, the Maxima code detects the minimum of the function given in given in Eq. (2.12) at the right-hand boundary, i.e $X_{extr} = 2.25$.

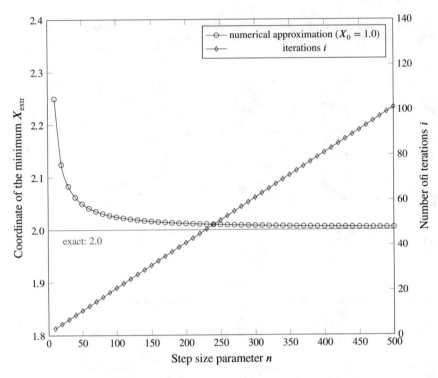

Fig. 2.8 Convergence rate of the brute-force approach (version 2) to detect the minimum of the function $F(X) = \frac{1}{2}(X - 2)^2 + 1$ in the range $0 \leq X \leq 5$

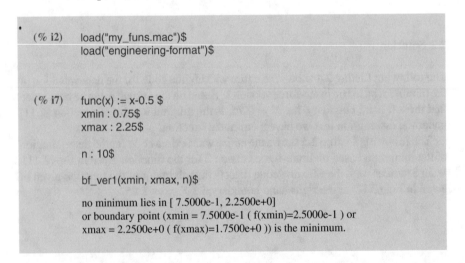

```
(% i2)    load("my_funs.mac")$
          load("engineering-format")$


(% i7)    func(x) := x-0.5 $
          xmin : 0.75$
          xmax : 2.25$

          n : 10$

          bf_ver1(xmin, xmax, n)$

          no minimum lies in [ 7.5000e-1, 2.2500e+0]
          or boundary point (xmin = 7.5000e-1 ( f(xmin)=2.5000e-1 ) or
          xmax = 2.2500e+0 ( f(xmax)=1.7500e+0 )) is the minimum.
```

Module 2.4: Numerical determination of the minimum for the function $F(X) = X - 0.5$ in the range $0.75 \leq X \leq 2.25$ based on the brute-force approach (version 1) for $n = 10$ (exact solution: $X_{extr} = 0.75$).

Fig. 2.9 Graphical representations of the objective functions according to **a** Eq. (2.11) and **b** Eq. (2.12). Exact solutions for the minima: $X_{\text{extr}} = 0.75$ and $X_{\text{extr}} = 2.25$, respectively

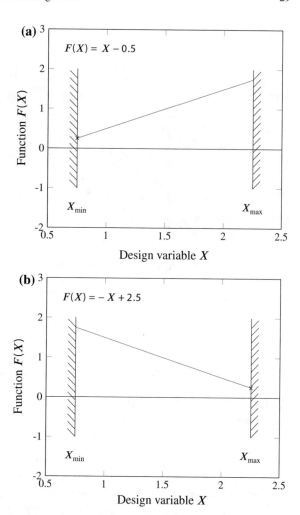

2.4 Numerical Determination of a Minimum Based on the Brute-Force Algorithm with Variable Interval Size

Determine based on the brute-force approach with variable step size the minimum of the quadratic function

$$F(X) = \tfrac{1}{2}(X - 2)^2 + 1 \tag{2.13}$$

in the range $0 \leq X \leq 5$ (see Fig. 2.2 for a graphical representation of the function). Assign to algorithm version 3 different scaling parameters $\alpha^{(i)}$ to update the interval size via $\Delta h^{(i)} = \alpha^{(i)} \times \Delta h^{(i-1)}$, i.e.

```
(% i2)      load("my_funs.mac")$
            load("engineering-format")$


(% i7)      func(x) := x-0.5 $
            xmin : 0.75$
            xmax : 2.25$
            x_0 : 1.0$

            n : 10$

            bf_ver2(xmin, xmax, x_0, n)$

            X_extr = 7.5000e-1 ( i = 2 )
```

Module 2.5: Numerical determination of the minimum for the function $F(X) = X - 0.5$ in the range $0.75 \leq X \leq 2.25$ based on the brute-force approach (version 2) for $n = 10$ and $X_0 = 1.0$ (exact solution: $X_{extr} = 0.75$).

- $\alpha^{(i)} = \frac{1}{10}$,
- $\alpha^{(i)} = 1.5$, and
- $\alpha^{(i)} = 1, 1, 2, 3, 5, 8, 13, \ldots (i > 0)$ (Fibonacci sequence).

Assign different start values ($X_0 = 1.0, 1.5, 3.5$) and parameters to control the initial step size ($n = 10, 25, 100, 1000$).

2.4 Solution

The following Listing 2.6 shows the entire wxMaxima code for the determination of the minimum based on brute-force version 3 for the function given in Eq. (2.13) and $\alpha^{(i)} = 1/10$. In case that the Fibonacci sequence should be used, the code must be modified by alpha: "Fibonacci".

Tables 2.3, 2.4 and 2.5 summarize the detected minimum for decreasing, increasing and changing interval sizes based on the Fibonacci sequence.

The convergence rate of the brute-force approach with variable step size based on the Fibonacci sequence (version 3) is illustrated in Fig. 2.10.

```
(% i2)    load("my_funs.mac")$
          load("engineering-format")$


(% i9)    func(x) := (1/2)*(x-2)^2 + 1$
          xmin : 0$
          xmax : 5$
          x0 : [1.0, 1.5, 3.5]$
          alpha : 1/10$
          n : [10, 25, 100, 1000]$
          for i : 1 thru length(x0) do (
              printf(true, "~% X_0 = ~f", x0[i]),
              for j : 1 thru length(n) do (
              bf_ver2_varN_table(xmin, xmax, x0[i], n[j], alpha)
              )
          )$



          X_0 = 1.0
              10 1.55555556e+0 17
              25 1.22222222e+0 17
             100 1.05555556e+0 16
            1000 1.00555556e+0 15
          X_0 = 1.5
              10 2.02500000e+0  2
              25 1.72222222e+0 16
             100 1.55555556e+0 16
            1000 1.50555556e+0 15
          X_0 = 3.5
              10 2.94444444e+0 17
              25 3.27777778e+0 16
             100 3.44444444e+0 16
            1000 3.49444444e+0 15
```

Module 2.6: Numerical determination of the minimum for the function $F(X) = \frac{1}{2}(X - 2)^2 + 1$ in the range $0 \leq X \leq 5$ based on the brute-force approach with variable step size (version 3) and $\alpha^{(i)} = 1/10$ (exact value: $X_{extr} = 2.0$).

2.3 Newton's Method

Newton's method is one of the classical second-order methods. These methods require the knowledge about the second-order derivative of the objective function, which must be provided as an analytical expression or approximation, for example,

Table 2.3 Summary of detected minimum values (exact value: $X_{extr} = 2.0$) for different initial values X_0 and parameters n, i.e. initial step sizes (brute-force version 3). Case: $\alpha^{(i)} = \frac{1}{10}$, i.e. decreasing interval size

n	X_{extr}	i
$X_0 = 1.0$		
10	1.55555556e+0	17
25	1.22222222e+0	17
100	1.05555556e+0	16
1000	1.00555556e+0	15
$X_0 = 1.5$		
10	2.02500000e+0	2
25	1.72222222e+0	16
100	1.55555556e+0	16
1000	1.50555556e+0	15
$X_0 = 3.5$		
10	2.94444444e+0	17
25	3.27777778e+0	16
100	3.44444444e+0	16
1000	3.49444444e+0	15

Table 2.4 Summary of detected minimum values (exact value: $X_{extr} = 2.0$) for different initial values X_0 and parameters n, i.e. initial step sizes (brute-force version 3). Case: $\alpha^{(i)} = 1.5$, i.e. increasing interval size

n	X_{extr}	i
$X_0 = 1.0$		
10	2.81250000e+0	3
25	2.28750000e+0	4
100	2.32382812e+0	7
1000	2.07121948e+0	12
$X_0 = 1.5$		
10	2.37500000e+0	2
25	2.22500000e+0	3
100	2.03281250e+0	5
1000	2.21081299e+0	11
$X_0 = 3.5$		
10	2.81250000e+0	3
25	2.38125000e+0	5
100	2.31855469e+0	8
1000	2.53690247e+0	13

Table 2.5 Summary of detected minimum values (exact value: $X_{extr} = 2.0$) for different initial values X_0 and parameters n, i.e. initial step sizes (brute-force version 3). Case: $\alpha^{(i)} = 1, 1, 2, 3, 5, 8, 13, \ldots$, i.e. increasing interval size based on Fibonacci sequence

n	X_{extr}	i
$X_0 = 1.0$		
10	2.25000000e+0	3
25	2.60000000e+0	5
100	2.30000000e+0	6
1000	2.77316000e+0	10
$X_0 = 1.5$		
10	2.25000000e+0	2
25	2.00000000e+0	3
100	2.80000000e+0	8
1000	2.30500000e+0	7
$X_0 = 3.5$		
10	2.50000000e+0	4
25	3.10000000e+0	5
100	2.35000000e+0	8
1000	2.52868000e+0	10

using finite difference schemes. Let us recall the necessary, i.e. $\frac{dF(X)}{dX} = 0$, and the sufficient, i.e. $\frac{d^2F(X)}{dX^2} > 0$, conditions for a relative minimum of the objective function $F(X)$.

Let us write a second-order Taylor series expansion of the objective function about X_0, i.e.

$$F(X) \approx F(X_0) + \frac{dF}{dX}\bigg|_{X_0} \times (X - X_0) + \frac{1}{2}\frac{d^2F}{dX^2}\bigg|_{X_0} \times (X - X_0)^2 . \qquad (2.14)$$

Differentiating Eq. (2.14) with respect to X and neglecting differences of higher-order gives finally the following expression considering the necessary condition for a local minimum:

$$\frac{dF(X)}{dX} \approx \frac{dF}{dX}\bigg|_{X_0} + \frac{d^2F}{dX^2}\bigg|_{X_0} \times (X - X_0) \stackrel{!}{=} 0 . \qquad (2.15)$$

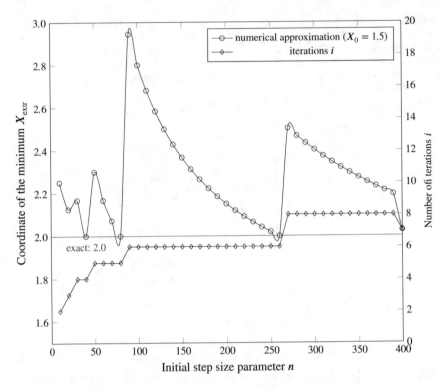

Fig. 2.10 Convergence rate of the brute-force approach with variable step size based on the Fibonacci sequence (version 3) to detect the minimum of the function $F(X) = \frac{1}{2}(X - 2)^2 + 1$ in the range $0 \leq X \leq 5$

The last expression can be rearranged to give the following iteration scheme[4]:

$$X = X_0 - \left(\frac{d^2 F}{dX^2}\bigg|_{X_0}\right)^{-1} \times \frac{dF}{dX}\bigg|_{X_0}. \qquad (2.16)$$

Possible finite difference approximations of the derivatives in Eq. (2.16) are given in Table 2.6. In the case that the computer algebra system Maxima is used, it is possible to calculate the derivatives based on built-in functions without the need for numerical approximations. Looking at Eq. (2.16), it can be concluded that the iteration scheme is not applicable for linear functions (i.e. first-order polynomials) since the second-order derivative would be zero and a division by zero is not defined. However, a linear function does not have any local minimum and only one of the boundaries,

---

[4]Let us remind the reader that Newton's method for the *root* of a function reads: $X = X_0 - \left(\frac{dF}{dX}\big|_{X_0}\right)^{-1} \times F(X_0)$.

Table 2.6 Finite difference approximations for first-and second-order derivatives, taken from [1]

Derivative	Finite difference approximation	Type	Error
$\left(\dfrac{dF}{dX}\right)_i$	$\dfrac{F_{i+1} - F_i}{\Delta X}$	Forward difference	$O(\Delta X)$
	$\dfrac{-3F_i + 4F_{i+1} - F_{i+2}}{2\Delta X}$	Forward difference	$O(\Delta X^2)$
	$\dfrac{F_i - F_{i-1}}{\Delta X}$	Backward difference	$O(\Delta X)$
	$\dfrac{3F_i - 4F_{i-1} + F_{i-2}}{2\Delta X}$	Backward difference	$O(\Delta X^2)$
	$\dfrac{F_{i+1} - F_{i-1}}{2\Delta X}$	Centered difference	$O(\Delta X^2)$
$\left(\dfrac{d^2 F}{dX^2}\right)_i$	$\dfrac{F_{i+2} - 2F_{i+1} + F_i}{\Delta X^2}$	Forward difference	$O(\Delta X)$
	$\dfrac{-F_{i+3} + 4F_{i+2} - 5F_{i+1} + 2F_i}{\Delta X^2}$	Forward difference	$O(\Delta X^2)$
	$\dfrac{F_i - 2F_{i-1} + F_{i-2}}{\Delta X^2}$	Backward difference	$O(\Delta X)$
	$\dfrac{2F_i - 5F_{i-1} + 4F_{i-2} - F_{i-3}}{\Delta X^2}$	Backward difference	$O(\Delta X^2)$
	$\dfrac{F_{i+1} - 2F_i + F_{i-1}}{\Delta X^2}$	Centered difference	$O(\Delta X^2)$

i.e., X_{min} or X_{max}, can be a minimum by definition. Thus, simply checking for the smaller value of $F(X_{min})$ and $F(X_{max})$ can be applied in the case of linear functions. An algorithmic realization of Newton's method using Eq. (2.16) is presented in Fig. 2.11.

2.5 Numerical Determination of a Minimum for an Unconstrained Function

Determine based on Newton's method the minimum of the function

$$F(X) = \tfrac{1}{2}(X - 2)^2 + 1 \tag{2.17}$$

in the range $0 \leq X \leq 5$ (see Fig. 2.2 for a graphical representation of the function). Assign different start values ($X_0 = 1.0, 1.5, 2.0$).

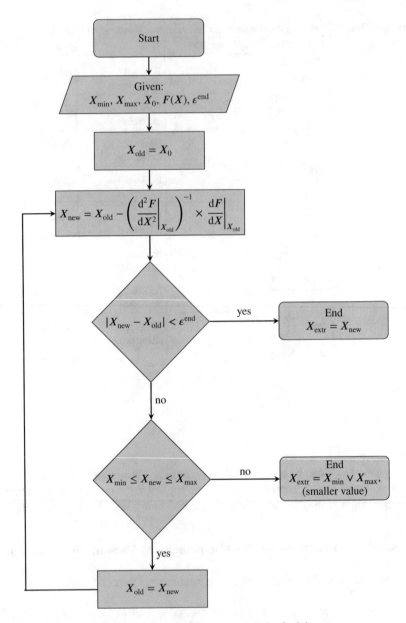

Fig. 2.11 Flowchart of the Newton's method for an unconstrained minimum

2.5 **Solution**

The following Listing 2.7 shows the entire wxMaxima code for the determination
of the minimum based on Newton's method. It can be seen that the minimum is
immediately found since function (2.17) is a second-order polynomial.

```
(% i2)    load("my_funs.mac")$
          load("engineering-format")$


(% i8)    func(X) := (1/2)*(X-2)^2 + 1$
          Xmin : 0$
          Xmax : 5$
          eps : 1/1000$
          X0s : [1,1.5,2]$

          for i : 1 thru length(X0s) do (
              print(" "),
              printf(true, "~%For X0 = ~,3f :", X0s[i]),
              Newton_one_variable_unconstrained(Xmin, Xmax, X0s[i], eps, true)
          )$



          For X0 = 1.000 :
            X_extr = 2.0000e+0 ( i = 2 )

          For X0 = 1.500 :
            X_extr = 2.0000e+0 ( i = 2 )

          For X0 = 2.000 :
            X_extr = 2.0000e+0 ( i = 1 )
```

Module 2.7: Numerical determination of the minimum for the functions (2.17), i.e.
$F(X) = \frac{1}{2}(X - 2)^2 + 1$, in the range $0 \leq X \leq 5$ based on Newton's method (exact
value: $X_{extr} = 2.0$).

2.6 **Numerical Determination of the Minimum for Two Unconstrained Functions**

Determine based on Newton's method the minimum of the functions

$$F(X) = X - 0.5 \tag{2.18}$$

and

$$F(X) = -X + 2.5 \qquad\qquad (2.19)$$

in the range $0.75 \leq X \leq 2.25$ (see Fig. 2.9 for graphical representations of the functions).

2.6 Solution

The following Listing 2.8 shows the entire wxMaxima code for the determination of the minimum based on Newton's method. Based on the output, it can be concluded that the left-hand boundary, i.e. $X = 0.75$, is the minimum point of function (2.18) under the assumption that we have a unimodal function. In a similar way, it can be concluded that the right-hand boundary, i.e. $X = 2.25$, is the minimum point of function (2.19).

```
(% i2)    load("my_funs.mac")$
          load("engineering-format")$


(% i8)    func(X) := -X + 2.5$
          /* func(X) := X - 0.5$ */
          Xmin : 0.75$
          Xmax : 2.25$
          eps : 1/1000$
          X0s : [1,1.5,2]$

          for i : 1 thru length(X0s) do (
              print(" "),
              printf(true, "~%For X0 = ~,3f :", X0s[i]),
              Newton_one_variable_unconstrained(Xmin, Xmax, X0s[i], eps, true)
          )$


          For X0 = 1.000 :
            X_extr = 2.2500e+0 ( i = 1 )

          For X0 = 1.500 :
            X_extr = 2.2500e+0 ( i = 1 )

          For X0 = 2.000 :
            X_extr = 2.2500e+0 ( i = 1 )
```

Module 2.8: Numerical determination of the minimum for the functions (2.18) and (2.19) in the range $0.75 \leq X \leq 2.25$ based on Newton's method (exact value: $X_{extr} = 2.25$).

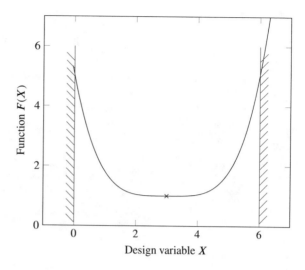

Fig. 2.12 Graphical representation of the objective function $F(X) = 0.05 \times (3 - X)^4 + 1$. Exact solution for the minimum: $X_{extr} = 3.0$

2.7 Numerical Determination of the Minimum for a Higher-Order Polynomial

Determine based on Newton's method the minimum of the function

$$F(X) = 0.05 \times (3 - X)^4 + 1 \tag{2.20}$$

in the range $0 \le X \le 6$ (see Fig. 2.12 for a graphical representation of the function). Assign different start values ($X_0 = 0.5, 1.0, 5.75$) for the iteration and take $\varepsilon^{end} = 1/1000$.

Investigate in the following the influence of the tolerance ε^{end} on the final value X_{extr} for the case $X_0 = 0.5$.

In a final step, investigate the development of the value X_{new} for a given X_0 and $\varepsilon^{end} = 1/1000$.

2.7 Solution

The following Listing 2.9 shows the entire wxMaxima code for the determination of the minimum based on Newton's method.

The following Listing 2.10 shows the entire wxMaxima code for the determination of the influence of the tolerance ε^{end} on the value of X_{extr}.

The following Listing 2.11 shows the entire wxMaxima code for the development of the value X_{new} for a given X_0 and $\varepsilon^{end} = 1/1000$. The graphical representation is given in Fig. 2.13.

```
(% i2)   load("my_funs.mac")$
         load("engineering-format")$


(% i8)   func(X) := 0.05*(3-X)^4 + 1$
         Xmin : 0$
         Xmax : 6$
         eps : 1/1000$
         X0s : [0.5,1.0,5.75]$

         for i : 1 thru length(X0s) do (
             print(" "),
             printf(true, "~%For X0 = ~,3f :", X0s[i]),
             Newton_one_variable_unconstrained(Xmin, Xmax, X0s[i], eps, true)
         )$


         For X0 = 0.500 :
           X_extr = 2.9983e+0 ( i = 18 )

         For X0 = 1.000 :
           X_extr = 2.9986e+0 ( i = 18 )

         For X0 = 5.750 :
           X_extr = 3.0019e+0 ( i = 18 )
```

Module 2.9: Numerical determination of the minimum for the function (1.18) in the range $0 \leq X \leq 6$ based on Newton's method (exact value: $X_{extr} = 3.0$).

```
(% i1)   load("my_funs.mac")$


(% i9)   func(X) := 0.05*(3-X)^4 + 1$
         Xmin : 0$
         Xmax : 6$
         eps : [1/10, 1/20,1/30,1/40,1/50,1/100,1/1000,1/10000,1/100000]$
         X0s : 0.5$

         print(" ")$
         printf(true, "~%~a ~a ~a ", "eps", "n_iter", "X_extr")$
         for i : 1 thru length(eps_s) do (
             [n_iter, X_extremum] : Newton_one_variable_unconstrained(Xmin,
             Xmax, X0, eps_s[i], false),
             printf(true, "~%~,5f ~,5d ~,8f ", eps_s[i], n_iter, X_extremum)
         )$
```

```
eps    n_iter  X_extr
0.10000  7    2.85368084
0.05000  8    2.90245389
0.03333  9    2.93496926
0.02500  10   2.95664618
0.02000  11   2.97109745
0.01000  12   2.98073163
0.00100  18   2.99830840
0.00010  24   2.99985149
0.00001  29   2.99998044
```

Module 2.10: Numerical determination of the minimum for the function (2.20) in the range $0 \leq X \leq 6$ based on Newton's method: influence of the tolerance ε^{end} on the final value X_{extr} (exact value: $X_{extr} = 3.0$)

```
(% i1)   load("my_funs.mac")$


(% i7)   func(X) := 0.05*(3-X)^4 + 1$
         Xmin : 0$
         Xmax : 6$
         eps : 1/1000$
         X0s : [0.5,1.0,5.75]$

         for i : 1 thru length(X0s) do (
             print(" "),
             print(" "),
             printf(true, "~%For X0 = ~,3f :", X0s[i]),
             print(" "),
             printf(true, "~%~a ~a ~a ", "iter", "Xnew", "func(X)"),
             Newton_one_variable_unconstrained_table(Xmin, Xmax, X0s[i], eps)
         )$


         For X0 = 0.500 :
         iter  Xnew      func(X)
          0  0.50000000  2.95312500
          1  1.33333333  1.38580247
          2  1.88888889  1.07620790
          3  2.25925926  1.01505341
          4  2.50617284  1.00297351
          5  2.67078189  1.00058736
          6  2.78052126  1.00011602
          7  2.85368084  1.00002292
          8  2.90245389  1.00000453
          9  2.93496926  1.00000089
```

```
10  2.95664618  1.00000018
11  2.97109745  1.00000003
12  2.98073163  1.00000001
13  2.98715442  1.00000000
14  2.99143628  1.00000000
15  2.99429085  1.00000000
16  2.99619390  1.00000000
17  2.99746260  1.00000000
18  2.99830840  1.00000000

For X0 = 1.000 :
iter  Xnew  func(X)
 0  1.00000000  1.80000000
 1  1.66666667  1.15802469
 2  2.11111111  1.03121475
 3  2.40740741  1.00616588
 4  2.60493827  1.00121795
 5  2.73662551  1.00024058
 6  2.82441701  1.00004752
 7  2.88294467  1.00000939
 8  2.92196312  1.00000185
 9  2.94797541  1.00000037
10  2.96531694  1.00000007
11  2.97687796  1.00000001
12  2.98458531  1.00000000
13  2.98972354  1.00000000
14  2.99314903  1.00000000
15  2.99543268  1.00000000
16  2.99695512  1.00000000
17  2.99797008  1.00000000
18  2.99864672  1.00000000

For X0 = 5.750 :
iter  Xnew  func(X)
 0  5.75000000  3.85957031
 1  4.83333333  1.56485340
 2  4.22222222  1.11157598
 3  3.81481481  1.02203970
 4  3.54320988  1.00435352
 5  3.36213992  1.00085995
 6  3.24142661  1.00016987
 7  3.16095107  1.00003355
 8  3.10730072  1.00000663
 9  3.07153381  1.00000131
10  3.04768921  1.00000026
11  3.03179280  1.00000005
12  3.02119520  1.00000001
13  3.01413014  1.00000000
14  3.00942009  1.00000000
15  3.00628006  1.00000000
16  3.00418671  1.00000000
17  3.00279114  1.00000000
18  3.00186076  1.00000000
```

Module 2.11: Numerical determination of the minimum for the function (2.20) in the range $0 \leq X \leq 6$ based on Newton's method: development of the value X_{new} for a given X_0 and $\varepsilon^{\text{end}} = 1/1000$ (exact value: $X_{\text{extr}} = 2.0$)

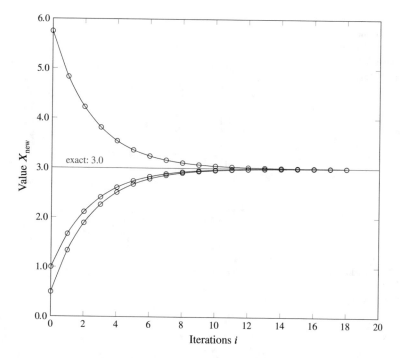

Fig. 2.13 Development of the value X_{new} for a given $X_0 (= 0.5, 1.0, 5.75)$ and $\varepsilon^{end} = 1/1000$

2.4 Supplementary Problems

2.8 Numerical Determination of a Minimum

Determine based on the golden section method the minimum of the function

$$F(X) = -2 \times \sin(X) \times (1 + \cos(X)) \tag{2.21}$$

in the range $0 \leq X \leq \frac{\pi}{2}$ (see Fig. 2.14). Assign for the relative tolerance the value $\varepsilon = 0.001$.

2.9 Numerical Determination of a Maximum

Determine based on the golden section method the maximum of the function

$$F(X) = \frac{8X}{X^2 - 2X + 4} \tag{2.22}$$

in the range $0 \leq X \leq 10$ (see Fig. 2.15). Assign for the relative tolerance the value $\varepsilon = 0.001$.

Fig. 2.14 Golden section method for the example $F(X) = -2 \times \sin(X) \times (1 + \cos(X))$

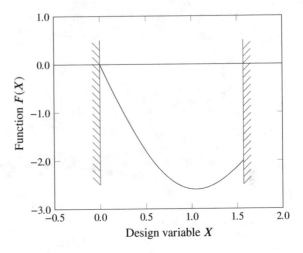

Fig. 2.15 Golden section method for the example $F(X) = \frac{8X}{X^2 - 2X + 4}$

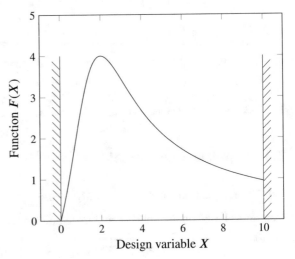

2.10 Brute-Force Approach for the Determination of a Minimum

Repeat supplementary problem 2.8 based on the two versions (i.e., version 1 and 2) of the brute-force approach. The step size should be increased in certain steps. Use as the start values $X_0 = 0.2$ and $X_0 = 1.3$ for version 2.

2.11 Convergence Rate for the Brute-Force Approach

Repeat supplementary problem 2.8 based on the second version of the brute-force approach to investigate the convergence rate. Plot the coordinate of the minimum as a function of the step size parameter n. Use as the start value $X_0 = 0.2$.

Fig. 2.16 Linear spring problem: **a** undeformed configuration; **b** elongated under the force F_0

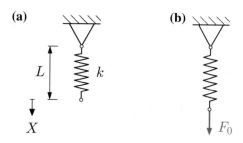

2.12 Numerical Determination of a Minimum Based on the Brute-Force Algorithm with Variable Interval Size

Determine based on the brute-force approach with variable step size the minimum of the function

$$F(X) = -2 \times \sin(X) \times (1 + \cos(X)) \qquad (2.23)$$

in the range $0 \leq X \leq \frac{\pi}{2}$ (see Fig. 2.14 for a graphical representation of the function). Assign to algorithm version 3 different scaling parameters $\alpha^{(i)}$ to update the interval size via $\Delta h^{(i)} = \alpha^{(i)} \times \Delta h^{(i-1)}$, i.e.

- $\alpha^{(i)} = \frac{1}{10}$,
- $\alpha^{(i)} = 1.5$, and
- $\alpha^{(i)} = 1, 1, 2, 3, 5, 8, 13, \ldots (i > 0)$ (Fibonacci sequence).

Assign different start values ($X_0 = 0.2, 1.3$) and parameters to control the initial step size ($n = 10, 15, 20, 25, 30, 35, 40, 1000$).

2.13 Application of the Principle of Minimum Energy to a Linear Spring Problem

Given is a linear spring (spring constant: $k = 8 \frac{N}{cm}$; length: $L = 10$ cm), see Fig. 2.16a. Under an applied force of $F_0 = 5$ N, the spring elongates and this deformation is described by the coordinate X, see Fig. 2.16b. Use the minimum total potential energy principle to derive the equilibrium position. The numerical approach should be based on the golden section algorithm.

References

1. Öchsner A (2014) Elasto-plasticity of frame structure elements: modeling and simulation of rods and beams. Springer-Verlag, Berlin
2. Vanderplaats GN (1999) Numerical optimization techniques for engineering design. Vanderplaats Research & Development, Colorado Springs

Chapter 3
Constrained Functions of One Variable

Abstract This chapter introduces two classical methods for the numerical determination of the minimum of unimodal functions of one variable. The exterior and
the interior penalty function methods are described. Based on a pseudo-objective
function, the problem can be treated as an unconstrained problem as covered in the
previous chapter.

3.1 The Exterior Penalty Function Method

Let us explain the exterior penalty function method with the general example of an
objective function $F(X)$ which is constrained by two inequality constraints $g_j(X)$,
see [4] for details (Fig. 3.1):

$$F(X),\tag{3.1}$$
$$g_1(X) \leq 0,\tag{3.2}$$
$$g_2(X) \leq 0.\tag{3.3}$$

A common approach in solving such a constrained design problem is to formulate a
so-called pseudo-objective function[1] (see Fig. 3.2 for a schematic representation)

$$\Phi(X, r_p) = F(X) + r_p \times P(X),\tag{3.4}$$

---

[1] Alternatively denoted as penalized objective.

Fig. 3.1 General
configuration of an objective
function $F(X)$. Two
inequality constraints $g_1 \leq 0$
and $g_2 \leq 0$ limit the
available design space

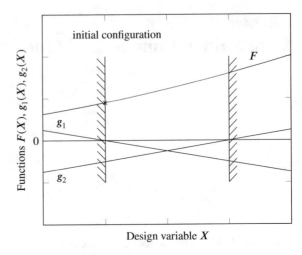

Fig. 3.2 Schematic
representation of the
pseudo-objective function
$\Phi(X, r_p)$ for different values
of the penalty multiplier r_p

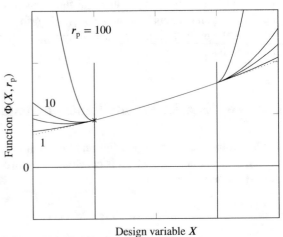

where $F(X)$ is the original objective function, r_p is the scalar (non-negative) penalty
parameter or penalty multiplier and $P(X)$ is the penalty function. The pseudo-
objective function can now be treated as an unconstrained function based on the
methods presented in Chap. 2, see Fig. 3.3 for the corresponding algorithm. The
penalty function is typically expressed as

$$P(X) = \sum_{j=1}^{m} \left\{ \max \left[0, g_j(X) \right] \right\}^2 , \tag{3.5}$$

Fig. 3.3 Algorithm for a constrained function of one variable based on the exterior penalty function method, adapted from [4]

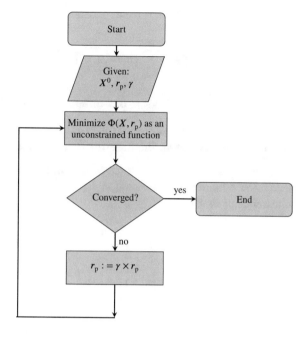

which can equally be expressed as

$$P(X) = \sum_{j=1}^{m} \delta_j \left\{ g_j(X) \right\}^2 . \tag{3.6}$$

Equation (3.6) uses the following step function:

$$\delta_j = \begin{cases} 0 \text{ for } g_j(X) \le 0 \\ 1 \text{ for } g_j(X) > 0 \end{cases} . \tag{3.7}$$

It should be noted here that in case of equality constraints (see Eq. (1.8)), the penalty function given in Eqs. (3.5) or (3.6) can be extended as follows:

$$P(X) = \sum_{j=1}^{m} \left\{ \max \left[0, g_j(X) \right] \right\}^2 + \sum_{k=1}^{l} [h_k(X)]^2 \tag{3.8}$$

$$= \sum_{j=1}^{m} \delta_j \left\{ g_j(X) \right\}^2 + \sum_{k=1}^{l} [h_k(X)]^2 , \tag{3.9}$$

where m and l are the total number of inequality and equality constraints, respectively.

An alternative approach is the application of linear penalty functions, i.e.

$$P(X) = \sum_{j=1}^{m} \{\max [0, g_j(X)]\}^1 + \sum_{k=1}^{l} [h_k(X)]^1 . \tag{3.10}$$

However, this approach leads to non-differentiable pseudo-objective functions.

3.1 Numerical Determination of the Minimum for a Function with Two Inequality Constraints Based on the Exterior Penalty Function Method

Determine based on the brute-force algorithm (version 3) for the one-dimensional search the minimum of the function $F(X)$ under consideration of two inequality constraints g_j in the range $0 \leq X \leq 5$:

$$F(X) = \frac{1}{2} \times (X - 2)^2 + 1 , \tag{3.11}$$

$$g_1(X) = \frac{1}{2} \times \left(\frac{1}{2} - X\right) \leq 0 , \tag{3.12}$$

$$g_2(X) = X - 4 \leq 0 . \tag{3.13}$$

The start value should be $X_0 = 1.5$ and the parameters for the one-dimensional search based on the brute-force algorithm are $\alpha^{(i)} = 1.0$ and $n = 10$. Assign for the scalar penalty parameter the values $r_p = 1, 10, 50, 100, 120$, and 600.

3.1 Solution

The pseudo-objective function can be written for this problem as

$$\Phi(X, r_p) = \frac{1}{2} \times (X - 2)^2 + 1 + r_p \Big\{ \left[\max \left(0, \tfrac{1}{2} \times \left(\tfrac{1}{2} - X\right)\right)\right]^2$$
$$+ [\max (0, X - 4)]^2 \Big\} , \tag{3.14}$$

or more explicitly for the different ranges:

$$X < \frac{1}{2} : \quad \Phi(X, r_p) = \frac{1}{2} \times (X - 2)^2 + 1 + r_p \left[\frac{1}{2} \times \left(\frac{1}{2} - X\right)\right]^2 , \tag{3.15}$$

$$\frac{1}{2} \leq X \leq 4 : \quad \Phi(X, r_p) = \frac{1}{2} \times (X - 2)^2 + 1 , \tag{3.16}$$

$$X > 4 : \quad \Phi(X, r_p) = \frac{1}{2} \times (X - 2)^2 + 1 + r_p [X - 4]^2 . \tag{3.17}$$

The following Listings 3.1 and 3.2 show the entire wxMaxima code for the determination of the minimum for the set of functions given in Eqs. (3.11)–(3.13).

```
(% i3)    load("my_funs.mac")$
          load(to_poly_solve)$ /* to check if all the roots are real (isreal_p(X))) */
          ratprint : false$


(% i22)   f(X) := (1/2)*(X-2)^2 + 1$
          g[1](X) := (1/2)*((1/2)-X)$
          g[2](X) := X-4$

          Xmin : 0$
          Xmax : 5$
          X0 : 1.5$
          alpha : 1$
          n : 10$

          r_p_list : [1, 10, 50, 100, 120, 600]$
          gamma : 1$

          print("===============================")$
          print("========== Solution ==========")$
          print("===============================")$
          print(" ")$
          print("The pseudo-objective function for different ranges of X:")$
          constrained_one_variable_range_detection()$
          print(" ")$
          print("===============================")$
          for i:1 thru length(r_p_list) do (
             r_p_0 : r_p_list[i],
             print(" "),
             printf(true, "~% For r_p = ~,6f :", r_p_0),
             print(" "),
             X_extr : one_variable_constrained_exterior_penalty(Xmin, Xmax, x0, n,
                alpha, r_p_0, gamma),
             print(" "),
             r_p : r_p_0,
             printf(true, "~% value of the non-penalized function at X = ~,6f : ~,6f", X_extr,
                              f(X_extr)),
             printf(true, "~% value of the penalized function at X = ~,6f : ~,6f", X_extr,
                              func(X_extr)),
             print(" "),
             print("===============================")
          )$
```

Module 3.1: Numerical determination of the minimum for the function (3.11) under consideration of the inequality constraints g_1 and g_2 in the range $0 \leq X \leq 5$ based on the exterior penalty function method

Fig. 3.4 **a** Representation of
the objective function $F(X)$
and the two inequality
constraints $g_1 \leq 0$ and
$g_2 \leq 0$. **b** Pseudo-objective
function $\Phi(X, r_\mathrm{p})$ for
different values of the
penalty multiplier r_p. Exact
solution for the minimum:
$X_\mathrm{extr} = 2.0$

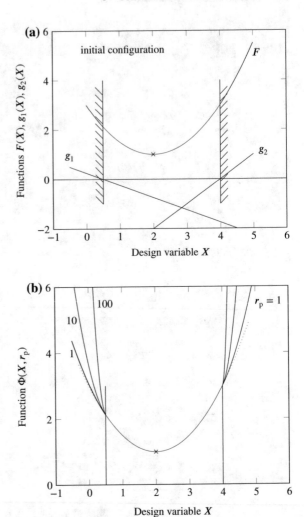

The graphical representation of the objective function and the two inequality constraints, as well as the pseudo-objective function for different values of the penalty
multiplier is given in Fig. 3.4.

```
===============================
========== Solution ==========
===============================
```

The pseudo-objective function for different ranges of X:

For 0.000000 < X < 0.500000 :

$$\varphi = \frac{(X-2)^2}{2} + \frac{r_p(\frac{1}{2}-X)^2}{4} + 1$$

For 0.500000 < X < 4.000000 :

$$\varphi = \frac{(X-2)^2}{2} + 1$$

For 4.000000 < X < 5.000000 :

$$\varphi = \frac{(X-2)^2}{2} + r_p(X-4)^2 + 1$$

```
===============================
```

For r_p = 1.000000 :

r_p: 1.000000 , X_extr: 2.250000 , Number of iterations: 2

value of the non-penalized function at X = 2.250000 : 1.031250
value of the penalized function at X = 2.250000 : 1.031250

```
===============================
```

For r_p = 10.000000 :

r_p: 10.000000 , X_extr: 2.500000 , Number of iterations: 1

value of the non-penalized function at X = 2.500000 : 1.125000
value of the penalized function at X = 2.500000 : 1.125000

```
===============================
```

For r_p = 50.000000 :

r_p: 50.000000 , X_extr: 2.250000 , Number of iterations: 2

value of the non-penalized function at X = 2.250000 : 1.031250
value of the penalized function at X = 2.250000 : 1.031250

==============================

For r_p = 100.000000 :

r_p: 100.000000 , X_extr: 2.500000 , Number of iterations: 1

value of the non-penalized function at X = 2.500000 : 1.125000
value of the penalized function at X = 2.500000 : 1.125000

==============================

For r_p = 120.000000 :

r_p: 120.000000 , X_extr: 2.250000 , Number of iterations: 2

value of the non-penalized function at X = 2.250000 : 1.031250
value of the penalized function at X = 2.250000 : 1.031250

==============================

For r_p = 600.000000 :

r_p: 600.000000 , X_extr: 2.500000 , Number of iterations: 2

value of the non-penalized function at X = 500000 : 1.125000
value of the penalized function at X = 2.500000 : 1.125000

==============================

Module 3.2: Numerical determination of the minimum for the function (3.11) under consideration of the inequality constraints g_1 and g_2 in the range $0 \leq X \leq 5$ based on the exterior penalty function method

3.2 Numerical Determination of the Minimum for a Function with One Inequality Constraint Based on the Exterior Penalty Function Method

Determine the minimum of the function $F(X)$ under consideration of the inequality constraint $g(X)$ in the range $-1 \leq X \leq 4$:

$$F(X) = (X - 3)^2 + 1,$$ (3.18)

$$g(X) = \sqrt{x} - \sqrt{2} \le 0.$$ (3.19)

The first approach should be based on the brute-force algorithm with $\alpha^{(i)} = 1.0$ and $n = 10000$. Assign for the scalar penalty parameter the values $r_p = 1, 10, 50, 100, 120$, and 600. As an alternative solution approach, use Newton's method to minimize $\Phi(X, r_p)$ as an unconstrained function ($\varepsilon^{end} = 1/1000$). The start value should be $X_0 = 1.0$ for both approaches.

3.2 Solution

The pseudo-objective function can be written for this problem as

$$\Phi(X, r_p) = (X - 3)^2 + 1 + r_p \left\{ \left[\max \left(0, \sqrt{x} - \sqrt{2} \right) \right]^2 \right\},$$ (3.20)

or more explicitly for the different ranges:

$$0 \le X \le 2: \quad \Phi(X, r_p) = (X - 3)^2 + 1,$$ (3.21)

$$X > 2: \quad \Phi(X, r_p) = (X - 3)^2 + 1 + r_p \left[\sqrt{x} - \sqrt{2} \right]^2.$$ (3.22)

The following Listing 3.3 shows the entire wxMaxima code for the determination of the minimum for the set of functions given in Eqs. (3.18) and (3.19) based on the brute-force algorithm (version 3).

As an alternative solution procedure, Listing 3.4 shows the entire wxMaxima code for the determination of the minimum for the set of functions given in Eqs. (3.18) and (3.19) based on Newton's method.

The graphical representation of the objective function and the inequality constraint, of the penalty function, and of the pseudo-objective function for different values of the penalty multiplier is given in Fig. 3.5.

Table 3.1 summarizes the results from Listings 3.3 and 3.4 in regard to iteration numbers and the obtained minima. It is quite obvious that Newton's method has a superb convergence speed.

3.3 Numerical Determination of the Minimum for a Constrained Higher-Order Polynomial Based on the Exterior Penalty Function Method

Determine based on Newton's method the minimum of the function

$$F(X) = 0.05 \times (3 - X)^4 + 1$$ (3.23)

Fig. 3.5 a Representation of
the objective function
$F(X) = (X − 3)^2 + 1$ and
the inequality constraint
$g \leq 0$. **b** Penalty function
$P(X)$ and inequality
constraint $g \leq 0$. **c**
Pseudo-objective function
$\Phi(X, r_p)$ for different values
of the penalty multiplier r_p.
Exact solution for the
minimum: $X_{extr} = 2.0$

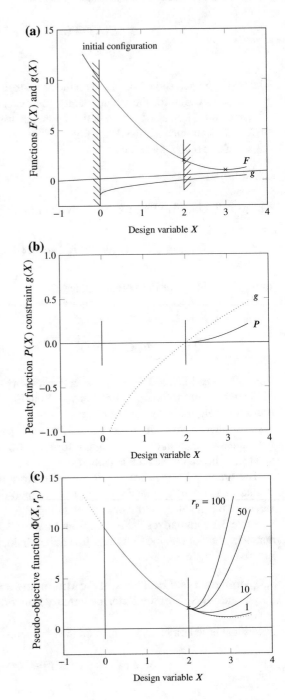

```
(% i3)    load("my_funs.mac")$
          load(to_poly_solve)$ /* to check if all the roots are real (isreal_p(X))) */
          ratprint : false$


(% i22)   f(X) := (X-3)^2 + 1$
          g[1](X) := sqrt(X) - sqrt(2)$

          Xmin : -1$
          Xmax : 4$
          X0 : 1$
          alpha : 1$
          n : 10000$

          r_p_list : [1, 10, 50, 100, 120, 600]$
          gamma : 1$

          print("===============================")$
          print("========== Solution ==========")$
          print("===============================")$
          print(" ")$
          print("The pseudo-objective function for different ranges of X:")$
          constrained_one_variable_range_detection()$
          print(" ")$
          print("===============================")$
          for i:1 thru length(r_p_list) do (
              r_p_0 : r_p_list[i],
              print(" "),
              printf(true, "~% For r_p = ~,6f :", r_p_0),
              print(" "),
              r_p : r_p_0,
              X_extr : one_variable_constrained_exterior_penalty(Xmin, Xmax, X0, n,
                  alpha, r_p_0, gamma),
              print(" "),
              printf(true, "~% value of the non-penalized function at X = ~,6f : ~,6f", X_extr,
                          f(X_extr)),
              printf(true, "~% value of the penalized function at X = ~,6f : ~,6f", X_extr,
                          func(X_extr)),
              print(" "),
              print("==============================")
          )$
```

```
===============================
========= Solution =========
===============================
```

The pseudo-objective function for different ranges of X:

For -1.000000 < X < 2.000000 :

$$\varphi = (X - 3)^2 + 1$$

For 2.000000 < X < 4.000000 :

$$\varphi = (X - 3)^2 + r_p\left(\sqrt{X} - \sqrt{2}\right)^2 + 1$$

```
===============================
```
For r_p = 1.000000 :

r_p: 1.000000 , X_extr: 2.917500 , Number of iterations: 284

value of the non-penalized function at X = 2.917500 : 1.006806
value of the penalized function at X = 2.917500 : 1.093157

```
===============================
```

For r_p = 10.000000 :

r_p: 10.000000 , X_extr: 2.490000 , Number of iterations: 87

value of the non-penalized function at X = 2.490000 : 1.260100
value of the penalized function at X = 2.490000 : 1.528273

```
===============================
```

For r_p = 50.000000 :

r_p: 50.000000 , X_extr: 2.147500 , Number of iterations: 70

value of the non-penalized function at X = 2.147500 : 1.726756
value of the penalized function at X = 2.147500 : 1.857938

```
===============================
```

For r_p = 100.000000 :

r_p: 100.000000 , X_extr: 2.080000 , Number of iterations: 15

value of the non-penalized function at X = 2.080000 : 1.846400
value of the penalized function at X = 2.080000 : 1.924839

==============================

For r_p = 120.000000 :

r_p: 120.000000 , X_extr: 2.067500 , Number of iterations: 4

value of the non-penalized function at X = 2.067500 : 1.869556
value of the penalized function at X = 2.067500 : 1.936770

==============================

For r_p = 600.000000 :

r_p: 600.000000 , X_extr: 2.015000 , Number of iterations: 12

value of the non-penalized function at X = 2.015000 : 1.970225
value of the penalized function at X = 2.015000 : 1.987037

==============================

Module 3.3: Numerical determination of the minimum for the function (3.18) under
consideration of the inequality constraint g in the range $-1 \leq X \leq 4$ based on the
exterior penalty function method and the brute-force algorithm (version 3)

```
(% i3)    load("my_funs.mac")$
          load(to_poly_solve)$ /* to check if all the roots are real (isreal_p(X))) */
          ratprint : false$


(% i22)   f(X) := (X-3)^2 + 1$
          g[1](X) := sqrt(X) - sqrt(2)$

          Xmin : -1$
          Xmax : 4$
          X0 : 1$
          eps : 1/1000$

          r_p_list : [1, 10, 50, 100, 120, 600]$
          gamma : 1$

          print("===============================")$
          print("========== Solution ==========")$
          print("===============================")$
```

```
print(" ")$
print("The pseudo-objective function for different ranges of X:")$
constrained_one_variable_range_detection()$
print(" ")$
print("===============================")$
for i:1 thru length(r_p_list) do (
    r_p_0 : r_p_list[i],
    print(" "),
    printf(true, "~% For r_p = ~,6f :", r_p_0),
    print(" "),
    r_p : r_p_0,
    X_extr : Newton_one_variable_constrained_exterior_penalty(Xmin, Xmax,
        X0, eps, r_p_0, gamma),
    X0 : copy(bfloat(X_extr)),
    print(" "),
    printf(true, "~% value of the non-penalized function at X = ~,6f : ~,6f", X_extr,
                 f(X_extr)),
    printf(true, "~% value of the penalized function at X = ~,6f : ~,6f", X_extr,
                 func(X_extr)),
    print(" "),
    print("===============================")
)$
```

```
===============================
========= Solution ==========
===============================
```

The pseudo-objective function for different ranges of X:

For $-1.000000 < X < 2.000000$:

$$\varphi = (X - 3)^2 + 1$$

For $2.000000 < X < 4.000000$:

$$\varphi = (X - 3)^2 + r_p \left(\sqrt{X} - \sqrt{2} \right)^2 + 1$$

```
===============================
```
For $r\_p = 1.000000$:

r_p: 1.000000 , X_extr: 2.914214 , Number of iterations: 3

value of the non-penalized function at X = 2.914214 : 1.007359
value of the penalized function at X = 2.914214 : 1.093146

==============================

For r_p = 10.000000 :

r_p: 10.000000 , X_extr: 2.485322 , Number of iterations: 3

value of the non-penalized function at X = 2.485322 : 1.264893
value of the penalized function at X = 2.485322 : 1.528231

==============================

For r_p = 50.000000 :

r_p: 50.000000 , X_extr: 2.144280 , Number of iterations: 3

value of the non-penalized function at X = 2.144280 : 1.732257
value of the penalized function at X = 2.144280 : 1.857870

==============================

For r_p = 100.000000 :

r_p: 100.000000 , X_extr: 2.076019 , Number of iterations: 3

value of the non-penalized function at X = 2.076019 : 1.853741
value of the penalized function at X = 2.076019 : 1.924636

==============================

For r_p = 120.000000 :

r_p: 120.000000 , X_extr: 2.063898 , Number of iterations: 2

value of the non-penalized function at X = 2.063898 : 1.876286
value of the penalized function at X = 2.063898 : 1.936572

==============================

For r_p = 600.000000 :

r_p: 600.000000 , X_extr: 2.013222 , Number of iterations: 2

value of the non-penalized function at X = 2.013222 : 1.973731
value of the penalized function at X = 2.013222 : 1.986799

==============================

Module 3.4: Numerical determination of the minimum for the function (3.18) under consideration of the inequality constraint g in the range $-1 \leq X \leq 4$ based on the exterior penalty function method (Newton's method)

Table 3.1 Comparison of the iteration numbers and obtained minima for the set of functions given in Eqs. (3.18)–(3.19) based on the brute-force approach and Newton's method

r_p	Iterations		X_{extr} (exact value: 2.0)	
	Brute-force	Newton	Brute-force	Newton
1	284	3	2.917500	2.914214
10	87	3	2.490000	2.485322
50	70	3	2.147500	2.144280
100	15	3	2.080000	2.076019
120	4	2	2.067500	2.063898
600	12	2	2.015000	2.013222

Fig. 3.6 Graphical representation of the objective function $F(X) = 0.05 \times (3 - X)^4 + 1$ and the inequality constraints $g_1 \leq 0$ and $g_2 \leq 0$, see Eqs. (3.23)–(3.25). Exact solution for the minimum: $X_{extr} = 3.0$

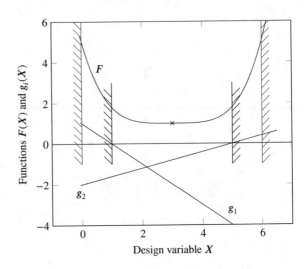

under consideration of the following inequality constraints:

$$g_1(X) = -X + 1 \leq 0, \tag{3.24}$$

$$g_2(X) = \frac{2}{5} \times X - 2 \leq 0, \tag{3.25}$$

in the range $0 \leq X \leq 6$ (see Fig. 3.6 for a graphical representation of the function). Assign different start values ($X_0 = 0.5, 1.5, 4.75$) for the iteration ($\varepsilon^{end} = 1/100000$).

3.3 **Solution**

The pseudo-objective function can be written for this problem as

$$\Phi(X, r_p) = 0.05 \times (3 - X)^4 + 1 + r_p \Big\{ [\max(0, -X + 1)]^2$$

$$+ \Big[\max \Big(0, \frac{2}{5} \times X - 2 \Big) \Big]^2 \Big\}, \quad (3.26)$$

or more explicitly for the different ranges:

$$0 \le X < 1: \quad \Phi(X, r_p) = 0.05 \times (3 - X)^4 + 1 + r_p [-X + 1]^2, \quad (3.27)$$
$$1 \le X \le 5: \quad \Phi(X, r_p) = 0.05 \times (3 - X)^4 + 1, \quad (3.28)$$

$$5 < X \le 6: \quad \Phi(X, r_p) = 0.05 \times (3 - X)^4 + 1 + r_p \Big[\frac{2}{5} \times X - 2 \Big]^2. \quad (3.29)$$

The following Listing 3.5 shows the entire wxMaxima code for the determination of the minimum for the set of functions given in Eqs. (3.23)–(3.25).

```
(% i3)    load("my_funs.mac")$
          load(to_poly_solve)$ /* to check if all the roots are real (isreal_p(X))) */
          ratprint : false$


(% i21)   f(X) := 0.05*(3-X)^4 + 1$
          g[1](X) := -X + 1$
          g[2](X) := (2/5)*X - 2$

          Xmin : 0$
          Xmax : 6$
          X0s : [0.5, 1.5, 4.75]$
          eps : 1/100000$

          r_p_list : [1, 10, 50, 100, 120, 600]$
          gamma : 1$

          print("==================================")$
          print("========== Solution ==========")$
          print("==================================")$
          print(" ")$
          print("The pseudo-objective function for different ranges of X:")$
          constrained_one_variable_range_detection()$
          print(" ")$
          print("==================================")$
```

```
for j : 1 thru length(X0s) do (
  X0 : copy(X0s[j]),
  print("+++++++++++++++++++++++++++++++++"),
  printf(true, "For X0 = ~,2f", X0),
  print("+++++++++++++++++++++++++++++++++"),
  for i:1 thru length(r_p_list) do (
    r_p_0 : r_p_list[i],
    print(" "),
    printf(true, "~% For r_p = ~,6f :", r_p_0),
    print(" "),
    r_p : r_p_0,
    X_extr : Newton_one_variable_constrained_exterior_penalty(Xmin, Xmax,
      X0, eps, r_p_0, gamma),
    X0 : copy(bfloat(X_extr)),
    r_p : r_p_0,
    print(" "),
    printf(true, "~% value of the non-penalized function at X = ~,6f : ~,6f", X_extr,
            f(X_extr)),
    printf(true, "~% value of the penalized function at X = ~,6f : ~,6f", X_extr,
            func(X_extr)),
    print(" "),
    print("==============================")
  )
)$
```

```
===============================
========== Solution ==========
===============================
```

The pseudo-objective function for different ranges of X:

For $0.000000 < X < 1.000000$:

$$\varphi = (1 - X)^2 r_p + 0.05(3 - X)^4 + 1$$

For $1.000000 < X < 5.000000$:

$$\varphi = 0.05(3 - X)^4 + 1$$

For $5.000000 < X < 6.000000$:

$$\varphi = \left(\frac{2X}{5} - 2\right)^2 r_p + 0.05(3 - X)^4 + 1$$

```
===============================
+++++++++++++++++++++++++++++++++
For X0 = 0.50
+++++++++++++++++++++++++++++++++
For r_p = 1.000000 :

r_p: 1.000000 , X_extr: 2.999986 , Number of iterations: 30

value of the non-penalized function at X = 2.999986 : 1.000000
value of the penalized function at X = 2.999986 : 1.000000


===============================

For r_p = 10.000000 :

r_p: 10.000000 , X_extr: 2.999991 , Number of iterations: 1

value of the non-penalized function at X = 2.2.999991 : 1.000000
value of the penalized function at X = 2.2.999991 : 1.000000


===============================

For r_p = 50.000000 :

r_p: 50.000000 , X_extr: 2.999994 , Number of iterations: 1

value of the non-penalized function at X = 2.999994 : 1.000000
value of the penalized function at X = 2.999994 : 1.000000


===============================

For r_p = 100.000000 :

r_p: 100.000000 , X_extr: 2.999996 , Number of iterations: 1

value of the non-penalized function at X = 2.999996 : 1.000000
value of the penalized function at X = 2.999996 : 1.000000


===============================

For r_p = 120.000000 :

r_p: 120.000000 , X_extr: 2.999997 , Number of iterations: 1

value of the non-penalized function at X = 2.999997 : 1.000000
value of the penalized function at X = 2.999997 : 1.000000


===============================

For r_p = 600.000000 :

r_p: 600.000000 , X_extr: 2.999998 , Number of iterations: 1

value of the non-penalized function at X = 2.999998 : 1.000000
value of the penalized function at X = 2.999998 : 1.000000


===============================
```

```
==============================
++++++++++++++++++++++++++++++
For X0 = 1.50
++++++++++++++++++++++++++++++
For r_p = 1.000000 :

r_p: 1.000000 , X_extr: 2.999982 , Number of iterations: 28

value of the non-penalized function at X = 2.999982 : 1.000000
value of the penalized function at X = 2.999982 : 1.000000


==============================


For r_p = 10.000000 :

r_p: 10.000000 , X_extr: 2.999988 , Number of iterations: 1

value of the non-penalized function at X = 2.2.999988 : 1.000000
value of the penalized function at X = 2.2.999988 : 1.000000


==============================


For r_p = 50.000000 :

r_p: 50.000000 , X_extr: 2.999992 , Number of iterations: 1

value of the non-penalized function at X = 2.999992 : 1.000000
value of the penalized function at X = 2.999992 : 1.000000


==============================


For r_p = 100.000000 :

r_p: 100.000000 , X_extr: 2.999995 , Number of iterations: 1

value of the non-penalized function at X = 2.999995 : 1.000000
value of the penalized function at X = 2.999995 : 1.000000


==============================


For r_p = 120.000000 :

r_p: 120.000000 , X_extr: 2.999997 , Number of iterations: 1

value of the non-penalized function at X = 2.999997 : 1.000000
value of the penalized function at X = 2.999997 : 1.000000


For r_p = 600.000000 :

r_p: 600.000000 , X_extr: 2.999998 , Number of iterations: 1

value of the non-penalized function at X = 2.999998 : 1.000000
value of the penalized function at X = 2.999998 : 1.000000
==============================
```

```
============================
++++++++++++++++++++++++++++++++
For X0 = 4.75
++++++++++++++++++++++++++++++++
For r_p = 1.000000 :

r_p: 1.000000 , X_extr: 3.000014 , Number of iterations: 29

value of the non-penalized function at X = 3.000014 : 1.000000
value of the penalized function at X = 3.000014 : 1.000000


============================


For r_p = 10.000000 :

r_p: 10.000000 , X_extr: 3.000009 , Number of iterations: 1

value of the non-penalized function at X = 3.000009 : 1.000000
value of the penalized function at X = 3.000009 : 1.000000


============================


For r_p = 50.000000 :

r_p: 50.000000 , X_extr: 3.000006 , Number of iterations: 1

value of the non-penalized function at X = 3.000006 : 1.000000
value of the penalized function at X = 3.000006 : 1.000000


============================


For r_p = 100.000000 :

r_p: 100.000000 , X_extr: 3.000004 , Number of iterations: 1

value of the non-penalized function at X = 3.000004 : 1.000000
value of the penalized function at X = 3.000004 : 1.000000


============================


For r_p = 120.000000 :

r_p: 120.000000 , X_extr: 3.000003 , Number of iterations: 1

value of the non-penalized function at X = 3.000003 : 1.000000
value of the penalized function at X = 3.000003 : 1.000000


============================
```

For r_p = 600.000000 :

r_p: 600.000000 , X_extr: 3.000002 , Number of iterations: 1

value of the non-penalized function at X = 3.000002 : 1.000000
value of the penalized function at X = 3.000002 : 1.000000

================================

Module 3.5: Numerical determination of the minimum for the function (3.23 under consideration of the inequality constraints g_1 and g_2 in the range $0 \leq X \leq 6$ based on the exterior penalty function method (Newton's method)

3.2 The Interior Penalty Function Method

The interior penalty function method considers the inequality conditions $g_j(X)$ in a different way than in the previous section. A common approach is to postulate the penalty function[2] $P(X)$ as [4]

$$P(X) = \sum_{j=1}^{m} \frac{-1}{g_j(X)} ,$$ (3.30)

or[3]

$$P(X) = \sum_{j=1}^{m} - \ln\left[-g_j(X)\right] .$$ (3.31)

Thus, the pseudo-penalty function $\Phi(X)$ in its general formulation, i.e. under the consideration of inequality (g_j) and equality (h_k) conditions, can be stated as follows:

$$\Phi(X, r'_p, r_p) = F(X) + r'_p \times P(X) + r_p \times \sum_{k=1}^{l} [h_k(X)]^2 ,$$ (3.32)

---

[2] In the context of the interior penalty function, $P(X)$ is also called the barrier function. Then, the penalty parameter r'_p is called the barrier parameter.

[3] Note that the natural logarithm is sometimes written as $\ln(x) = \log(x)$, i.e. the 'log' function without any subscript.

where r'_p is the penalty parameter for the inequality conditions and r_p is the penalty parameter for the equality conditions. During the algorithmic iteration, r'_p is decreased whereas r_p is increased as in the case of the exterior penalty method.[4]

Let us schematically illustrate the course of the pseudo-penalty function for the simplified case that the objective function $F(X)$ is only constrained by two inequality constraints $g_j(X)$, see Fig. 3.7a:

$$F(X),\tag{3.33}$$
$$g_1(X) \le 0,\tag{3.34}$$
$$g_2(X) \le 0.\tag{3.35}$$

Considering in addition that the penalty function is given by Eq. (3.30), the pseudo-penalty function reads, see Fig. 3.7b:

$$\Phi(X, r'_\mathrm{p}) = F(X) + r'_\mathrm{p} \times \left(-\frac{1}{g_1(X)} - \frac{1}{g_2(X)} \right).\tag{3.36}$$

The algorithm for the interior penalty function method is basically the same as for the exterior penalty method, see Fig. 3.3. However, the difference is that the penalty parameter r'_p for the inequality conditions $g_j(X)$ is decreased for the interior penalty method.

3.4 Numerical Determination of the Minimum for a Function with Two Inequality Constraints Based on the Interior Penalty Function Method

Determine based on the brute-force algorithm (version 3) for the one-dimensional search the minimum of the function $F(X)$ under consideration of two inequality constraints g_j in the range $0 \le X \le 5$ (see Fig. 3.8a for a graphical representation of $F(X)$ and g_j):

$$F(X) = \frac{1}{2} \times (X - 2)^2 + 1,\tag{3.37}$$

$$g_1(X) = \frac{1}{2} \times \left(\frac{1}{2} - X \right) \le 0,\tag{3.38}$$

$$g_2(X) = X - 4 \le 0.\tag{3.39}$$

The start value should be $X_0 = 3.0$ and the parameters for the one-dimensional search based on the brute-force algorithm are $\alpha^{(i)} = 1.0$ and $n = 1000$. Assign for the scalar penalty parameter the values $r'_\mathrm{p} = 0.5, 0.4, 0.3, 0.2, 0.1$ and 0.01.

---

[4]The treatment of equality constraints h_k is for both approaches, i.e., the interior and the exterior penalty function method, the same, see Eqs. (3.32) and (3.8).

Fig. 3.7 a General configuration of a constrained function $F(X)$. The two inequality constraints $g_1 \leq 0$ and $g_2 \leq 0$ limit the available design space. **b** Pseudo-objective function $\Phi(X)$ for different values of the penalty multiplier r'_p

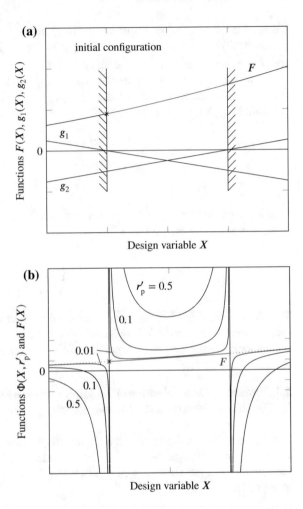

3.4 **Solution**

The pseudo-objective function can be written for this problem based on the fractional formulation (see Eq. (3.30)) as

$$\Phi(X, r'_\mathrm{p}) = \frac{1}{2} \times (X - 2)^2 + 1 + r'_\mathrm{p} \left\{ -\frac{1}{\frac{1}{2}\left(\frac{1}{2} - X\right)} - \frac{1}{X - 4} \right\}, \tag{3.40}$$

or alternatively in a logarithmic formulation (see Eq. (3.31))

$$\Phi(X, r'_\mathrm{p}) = \frac{1}{2} \times (X - 2)^2 + 1 + r'_\mathrm{p} \left\{ -\ln\left(-\frac{1}{2}\left(\frac{1}{2} - X\right)\right) - \ln\left(-(X - 4)\right) \right\}. \tag{3.41}$$

Fig. 3.8 a Representation of the objective function $F(X)$ (in blue color) and the two inequality constraints $g_1 \leq 0$ and $g_2 \leq 0$. **b** Pseudo-objective function $\Phi(X)$ for different values of the penalty multiplier $r'_p = 0.5, 0.1, 0.01$ based on the fractional representation. Exact solution for the minimum: $X_{\text{extr}} = 2.0$

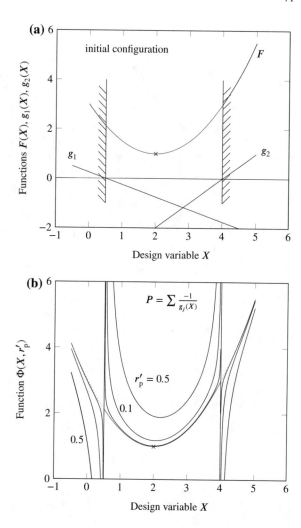

Based on the domain of the $\ln(x)$ function, i.e. $\mathbb{R}^+$, the pseudo-penalty function in Eq. (3.41) is only defined in the range $\frac{1}{2} \leq X \leq 4$, see Fig. 3.9.

The following Listing 3.6 shows the entire wxMaxima code for the determination of the minimum for the set of functions given in Eqs. (3.37)–(3.39) based on the approaches (3.40) and (3.41).

```
(% i3)    load("my_funs.mac")$
          load(to_poly_solve)$ /* to check if all the roots are real (isreal_p(X))) */
          ratprint : false$




(% i32)   f(X) := (1/2)*(X-2)^2 + 1$
          g[1](X) := (1/2)*((1/2)-X)$
          g[2](X) := X-4$

          Xmin : -1$
          Xmax : 6$
          X0 : 3$
          alpha : 1$
          n : 1000$

          r_p : [0]$
          r_p_prime_list : [0.5, 0.4, 0.3, 0.2, 0.1, 0.01]$
          gamma : 1$

          print("===============================")$
          print("========== Solution ==========")$
          print("===============================")$
          print(" ")$
          print("The pseudo-objective function:")$
          print(" ")$
          print("Fractional")$
          pseudo_function_interior_penalty("Fractional")$
          print(%Phi,"=",func(X))$
          print(" ")$
          print("Logarithmic")$
          pseudo_function_interior_penalty("Logarithmic")$
          print(%Phi,"=",func(X))$
          print(" ")$
          print("===============================")$
          X0_frac : X0$
          X0_log : X0$
          for i:1 thru length(r_p_prime_list) do (
              r_p_0 : r_p[1],
              r_p_prime_0 : r_p_prime_list[i],
              print(" "),
              printf(true, "~%For r_p_prime = ~,6f :", r_p_prime_0),
              print(" "),
              printf(true, "~%Fractional penalty function:"),
              type : "Fractional",
```

```
X_extr_frac : constrained_one_variable_interior_penalty(Xmin, Xmax, X0_frac,
                n, alpha, r_p_0, r_p_prime_0, gamma, type),
printf(true, "~% value of the non-penalized function at X = ~,6f : ~,6f",
                X_extr_frac, f(X_extr_frac)),
printf(true, "~% value of the penalized function at X = ~,6f : ~,6f",
                X_extr_frac, func(X_extr_frac)),
print(" "),
printf(true, "~%Logarithmic penalty function:"),
type :Ã,Â "Logarithmic",
X_extr_log : one_variable_constrained_interior_penalty(Xmin, Xmax, X0_log,
                n, alpha, r_p_0, r_p_prime_0, gamma, type),
printf(true, "~% value of the non-penalized function at X = ~,6f : ~,6f",
                X_extr_log, f(X_extr_log)),
printf(true, "~% value of the penalized function at X = ~,6f : ~,6f",
                X_extr_log, func(X_extr_log)),
print(" "),
print("=============================="),
X0_frac : X_extr_frac,
X0_log : X_extr_log
)$
```

```
==============================
========= Solution ==========
==============================
```

The pseudo-objective function:

Fractional

$$\Phi = \left(-\frac{1}{X-4} - \frac{2}{\frac{1}{2}-X} \right) r\_p\_prime + \frac{(X-2)^2}{2} + 1$$

Logarithmic

$$\Phi = \left(-\log(4-X) - \log\left(-\frac{\frac{1}{2}-X}{2} \right) \right) r\_p\_prime + \frac{(X-2)^2}{2} + 1$$

```
==============================
```

For r_p_prime = 0.500000 :

Fractional penalty function:
X_0: 3.000000 , r_p_prime: 0.500000 , X_extr: 2.198500 , Number of iterations: 116
value of the non-penalized function at X = 2.198500 : 1.019701
value of the penalized function at X = 2.198500 : 1.886002

Logarithmic penalty function:
X_0: 3.000000 , r_p_prime: 0.500000 , X_extr: 2.065500 , Number of iterations: 135
value of the non-penalized function at X = 2.065500 : 1.002145
value of the penalized function at X = 2.065500 : 0.794692

================================

For r_p_prime = 0.400000 :

Fractional penalty function:
X_0: 2.198500 , r_p_prime: 0.400000 , X_extr: 2.174000 , Number of iterations: 5
value of the non-penalized function at X = 2.174000 : 1.015138
value of the penalized function at X = 2.174000 : 1.712093

Logarithmic penalty function:
X_0: 2.065500 , r_p_prime: 0.400000 , X_extr: 2.055000 , Number of iterations: 3
value of the non-penalized function at X = 2.055000 : 1.001513
value of the penalized function at X = 2.055000 : 0.836076

================================

For r_p_prime = 0.300000 :

Fractional penalty function:
X_0: 2.174000 , r_p_prime: 0.300000 , X_extr: 2.142500 , Number of iterations: 6
value of the non-penalized function at X = 2.142500 : 1.010153
value of the penalized function at X = 2.142500 : 1.536957

Logarithmic penalty function:
X_0: 2.055000 , r_p_prime: 0.300000 , X_extr: 2.044500 , Number of iterations: 3
value of the non-penalized function at X = 2.044500 : 1.000990
value of the penalized function at X = 2.044500 : 0.877330

================================

For r_p_prime = 0.200000 :

Fractional penalty function:
X_0: 2.142500 , r_p_prime: 0.200000 , X_extr: 2.104000 , Number of iterations: 7
value of the non-penalized function at X = 2.104000 : 1.005408
value of the penalized function at X = 2.104000 : 1.360270

Logarithmic penalty function:
X_0: 2.044500 , r_p_prime: 0.200000 , X_extr: 2.034000 , Number of iterations: 3
value of the non-penalized function at X = 2.034000 : 1.000578
value of the penalized function at X = 2.034000 : 0.918431

================================

For r_p_prime = 0.100000 :

Fractional penalty function:
X_0: 2.104000 , r_p_prime: 0.100000 , X_extr: 2.058500 , Number of iterations: 8
value of the non-penalized function at X = 2.058500 : 1.001711
value of the penalized function at X = 2.058500 : 1.181546

Logarithmic penalty function:
X_0: 2.034000 , r_p_prime: 0.100000 , X_extr: 2.016500 , Number of iterations: 4
value of the non-penalized function at X = 2.016500 : 1.000136
value of the penalized function at X = 2.016500 : 0.959324

===============================

For r_p_prime = 0.010000 :

Fractional penalty function:
X_0: 2.058500 , r_p_prime: 0.010000 , X_extr: 2.013000 , Number of iterations: 8
value of the non-penalized function at X = 2.013000 : 1.000085
value of the penalized function at X = 2.013000 : 1.018336

Logarithmic penalty function:
X_0: 2.016500 , r_p_prime: 0.010000 , X_extr: 2.006000 , Number of iterations: 3
value of the non-penalized function at X = 2.006000 : 1.000018
value of the penalized function at X = 2.006000 : 0.995953

===============================

Module 3.6: Numerical determination of the minimum for the function (3.37 under consideration of the inequality constraints g_1 and g_2 in the range $-1 \leq X \leq 6$ based on the exterior penalty function method

Table 3.2 summarizes the results from Listing 3.6 in regard to iteration numbers and the obtained minima. It can be seen that the fractional and the logarithmic formulation of the penalty function result in quite similar results in regard to convergence speed and the detected minimum. However, an important conclusion from this example is that the choice of the start value (in our case: 3.0) had to be inside the interval $0.5 < X < 4$. Any choice outside this interval would have not detected the minimum ($X_{extr} = 2.0$) of the objective function.

3.5 Numerical Determination of the Minimum for a Function with One Inequality Constraint Based on the Interior Penalty Function Method

Determine based on the brute-force algorithm (version 3) for the one-dimensional search the minimum of the function $F(X)$ under consideration of one inequality constraint g in the range $-1 \leq X \leq 4$ (see Fig. 3.5a for a graphical representation of $F(X)$ and g):

Fig. 3.9 Pseudo-objective
function $\Phi(X)$ for different
values of the penalty
multiplier $r_p' = 0.5, 0.1, 0.01$
based on the logarithmic
formulation. Exact solution
for the minimum:
$X_{\text{extr}} = 2.0$

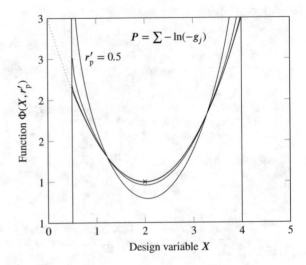

Table 3.2 Comparison of the iteration numbers and obtained minima for the set of functions given in Eqs. (3.37)–(3.39) based on the brute-force approach and Newton's method

r_p'	Iterations		X_{extr} (exact value: 2.0)	
	Fractional	Logarithmic	Fractional	Logarithmic
0.5	116	135	2.198500	2.065500
0.4	5	3	2.174000	2.055000
0.3	6	3	2.142500	2.044500
0.2	7	3	2.104000	2.034000
0.1	8	4	2.058500	2.016500
0.01	8	3	2.013000	2.006000

$$F(X) = (X - 3)^2 + 1 , \qquad (3.42)$$
$$g(X) = \sqrt{x} - \sqrt{2} \le 0 . \qquad (3.43)$$

The start value should be $X_0 = 1.0$ and the parameters for the one-dimensional search based on the brute-force algorithm are $\alpha^{(i)} = 1.0$ and $n = 1000$. Assign for the scalar penalty parameter the values $r_p' = 0.5, 0.4, 0.3, 0.2, 0.1$ and 0.01.

3.5 Solution

The pseudo-objective function (see Fig. 3.10) can be written for this problem based on the fractional formulation (see Eq. (3.30)) as

$$\Phi(X, r_p') = (X - 3)^2 + 1 + r_p' \left\{ -\frac{1}{\sqrt{x} - \sqrt{2}} \right\} , \qquad (3.44)$$

Fig. 3.10 Pseudo-objective function $\Phi(X, r'_p)$ for different values of the penalty multiplier $r'_p = 0.5, 0.1$: **a** $P = \frac{-1}{g(X)}$, **b** $P = -\ln(-g)$. Exact solution for the minimum: $X_{\text{extr}} = 2.0$

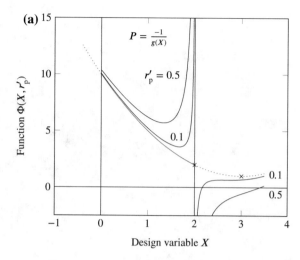

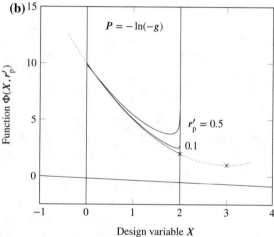

or alternatively in a logarithmic formulation (see Eq. (3.31))

$$\Phi(X, r'_p) = (X - 3)^2 + 1 + r'_p \left\{ -\ln\left(-\sqrt{x} + \sqrt{2}\right) \right\} . \qquad (3.45)$$

Based on the domain of the $\ln(x)$ function, i.e. $\mathbb{R}^+$, the pseudo-penalty function in Eq. (3.45) is only defined in the range $X \leq 2$, see Fig. 3.10b.

The following Listing 3.7 shows the entire wxMaxima code for the determination of the minimum for the set of functions given in Eqs. (3.42) and (3.43) based on the approaches (3.44) and (3.45).

Table 3.3 summarizes the results from Listing 3.7 in regard to iteration numbers and the obtained minima. It can be seen that the the logarithmic formulation of the

```
(% i3)    load("my_funs.mac")$
          load(to_poly_solve)$ /* to check if all the roots are real (isreal_p(X)) */
          ratprint : false$


(% i31)   f(X) := (X-3)^2 + 1$
          g[1](X) := sqrt(x)-sqrt(2)$



Xmin : -1$
Xmax : 4$
X0 : 1$
alpha : 1$
n : 1000$

r_p : [0]$
r_p_prime_list : [0.5, 0.4, 0.3, 0.2, 0.1, 0.01]$
gamma : 1$

print("================================")$
print("========== Solution ==========")$
print("================================")$
print(" ")$
print("The pseudo-objective function:")$
print(" ")$
print("Fractional")$
pseudo_function_interior_penalty("Fractional")$
print(%Phi,"=",func(X))$
print(" ")$
print("Logarithmic")$
pseudo_function_interior_penalty("Logarithmic")$
print(%Phi,"=",func(X))$
print(" ")$
print("================================")$
X0_frac : X0$
X0_log : X0$
for i:1 thru length(r_p_prime_list) do (
  r_p_0 : r_p[1],
  r_p_prime_0 : r_p_prime_list[i],
  print(" "),
  printf(true, "~%For r_p_prime = ~,6f :", r_p_prime_0),
  print(" "),
  printf(true, "~%Fractional penalty function:"),
  type :Ã,Â "Fractional",
  X_extr_frac : constrained_one_variable_interior_penalty(Xmin, Xmax, X0_frac,
                  n, alpha, r_p_0, r_p_prime_0, gamma, type),
  printf(true, "~% value of the non-penalized function at X = ~,6f : ~,6f",
                  X_extr_frac, f(X_extr_frac)),
  printf(true, "~% value of the penalized function at X = ~,6f : ~,6f",

                  X_extr_frac, func(X_extr_frac)),
```

```
  print(" "),
  printf(true, "~%Logarithmic penalty function:"),
  type : "Logarithmic",
  X_extr_log : one_variable_constrained_interior_penalty(Xmin, Xmax, X0_log,
                n, alpha, r_p_0, r_p_prime_0, gamma, type),
  printf(true, "~% value of the non-penalized function at X = ~,6f : ~,6f",
                X_extr_log, f(X_extr_log)),
  printf(true, "~% value of the penalized function at X = ~,6f : ~,6f",
                X_extr_log, func(X_extr_log)),
  print(" "),
  print("=============================="),
  X0_frac : X_extr_frac,
  X0_log : X_extr_log
)$
```

```
==============================
========== Solution ==========
==============================
```

The pseudo-objective function:

Fractional

$$\Phi = -\frac{r\_p\_prime}{\sqrt{X} - \sqrt{2}} + (X - 3)^2 + 1$$

Logarithmic

$$\Phi = -\log\left(\sqrt{2} - \sqrt{X}\right)r\_p\_prime + (X - 3)^2 + 1$$

```
==============================
```

For r_p_prime = 0.500000 :

Fractional penalty function:
X_0: 1.000000 , r_p_prime: 0.500000 , X_extr: 1.347500 , Number of iterations: 70
value of the non-penalized function at X = 1.347500 : 3.730756
value of the penalized function at X = 1.347500 : 5.703961

Logarithmic penalty function:
X_0: 1.000000 , r_p_prime: 0.500000 , X_extr: 1.792500 , Number of iterations: 159
value of the non-penalized function at X = 1.792500 : 2.458056
value of the penalized function at X = 1.792500 : 3.750724

```
==============================
```

For r_p_prime = 0.400000 :

Fractional penalty function:
X_0: 1.347500 , r_p_prime: 0.400000 , X_extr: 1.405000 , Number of iterations: 12
value of the non-penalized function at X = 1.405000 : 3.544025
value of the penalized function at X = 1.405000 : 5.291615

Logarithmic penalty function:
X_0: 1.792500 , r_p_prime: 0.400000 , X_extr: 1.830000 , Number of iterations: 8
value of the non-penalized function at X = 1.830000 : 2.368900
value of the penalized function at X = 1.830000 : 3.484787

===============================

For r_p_prime = 0.300000 :
Fractional penalty function:
X_0: 1.405000 , r_p_prime: 0.300000 , X_extr: 1.472500 , Number of iterations: 14
value of the non-penalized function at X = 1.472500 : 3.333256
value of the penalized function at X = 1.472500 : 4.827671

Logarithmic penalty function:
X_0: 1.830000 , r_p_prime: 0.300000 , X_extr: 1.867500 , Number of iterations: 8
value of the non-penalized function at X = 1.867500 : 2.282556
value of the penalized function at X = 1.867500 : 3.195727

===============================

For r_p_prime = 0.200000 :

Fractional penalty function:
X_0: 1.472500 , r_p_prime: 0.200000 , X_extr: 1.560000 , Number of iterations: 18
value of the non-penalized function at X = 1.560000 : 3.073600
value of the penalized function at X = 1.560000 : 4.284151

Logarithmic penalty function:
X_0: 1.867500 , r_p_prime: 0.200000 , X_extr: 1.910000 , Number of iterations: 9
value of the non-penalized function at X = 1.910000 : 2.188100
value of the penalized function at X = 1.910000 : 2.875344

===============================

For r_p_prime = 0.100000 :

Fractional penalty function:
X_0: 1.560000 , r_p_prime: 0.100000 , X_extr: 1.677500 , Number of iterations: 24
value of the non-penalized function at X = 1.677500 : 2.749006
value of the penalized function at X = 1.677500 : 3.589129

Logarithmic penalty function:
X_0: 1.910000 , r_p_prime: 0.100000 , X_extr: 1.952500 , Number of iterations: 9
value of the non-penalized function at X = 1.952500 : 2.097256
value of the penalized function at X = 1.952500 : 2.505332

```
===============================
For r_p_prime = 0.010000 :

Fractional penalty function:
X_0: 1.677500 , r_p_prime: 0.010000 , X_extr: 1.890000 , Number of iterations: 43
value of the non-penalized function at X = 1.890000 : 2.232100
value of the penalized function at X = 1.890000 : 2.485644
Logarithmic penalty function:
X_0: 1.952500 , r_p_prime: 0.010000 , X_extr: 1.995000 , Number of iterations: 9
value of the non-penalized function at X = 1.995000 : 2.010025
value of the penalized function at X = 1.995000 : 2.073399


===============================
```

Module 3.7: Numerical determination of the minimum for the function (3.42) under consideration of the inequality constraints g_1 and g_2 in the range $-1 \leq X \leq 4$ based on the exterior penalty function method

penalty function results in slightly better results in regard to convergence speed and the detected minimum compared to the fractional formulation. Again, the important conclusion from this example is that the choice of the start value (in our case: 1.0) had to be inside the interval $0 < X < 2$. Any choice outside this interval would hardly detected the minimum ($X_{\text{extr}} = 2.0$) of the objective function.

Table 3.3 Comparison of the iteration numbers and obtained minima for the set of functions given in Eqs. (3.42) and (3.43) based on the brute-force approach and Newton's method

r'_p	Iterations		X_{extr} (exact value: 2.0)	
	Fractional	Logarithmic	Fractional	Logarithmic
0.5	70	159	1.347500	1.792500
0.4	12	8	1.405000	1.830000
0.3	14	8	1.472500	1.867500
0.2	18	9	1.560000	1.910000
0.1	24	9	1.677500	1.952500
0.01	43	9	1.890000	1.99500

3.3   Supplementary Problems

3.6   Numerical Determination of the Optimal Design of a Cantilever Beam

Given is a cantilever beam as shown in Fig. 3.11. The beam is loaded by a single force F_0 and has constant material (E, ϱ) and geometrical properties (I) along its axis. The material is isotropic and homogeneous and the beam theory for thin beams (Euler-Bernoulli) should be applied for this example, see [2].

Given are:

- Geometrical dimension: $L = 2540\,\text{mm}$.
- Material properties of the beam: Young's modulus $E = 68948\,\text{MPa}$, mass density $\varrho = 2691\,\text{kg/m}^3$, initial tensile yield stress $R_{p0.2} = 247\,\text{MPa}$.
- Loading: $F_0 = 2667\,\text{N}$.

Determine the optimized cross-sectional dimension a under the condition that the acting stress does not exceed the initial yield stress. Furthermore, the beam should not exceed a maximum deflection of $u_z(L) = r_1 L$ with $r_1 = 0.03$. Use the exterior penalty function method to solve this problem.

3.7   Numerical Determination of the Optimal Design of a Compression Strut

Given is a compression strut as shown in Fig. 3.12. The rod is loaded by a single force F_0 and has constant material (E, ϱ) and geometrical properties (A) along its axis. Furthermore, the material is isotropic and homogeneous.

Given are:

- Geometrical dimension: $L = 1000\,\text{mm}$.
- Material properties of the rod: Young's modulus $E = 70000\,\text{MPa}$, mass density $\varrho = 2691\,\text{kg/m}^3$, initial tensile yield stress $R_{p0.2} = 247\,\text{MPa}$.
- Loading: $F_0 = 2667\,\text{N}$.

Determine the optimized cross-sectional dimension a under the condition that the acting stress does not exceed the initial yield stress. Furthermore, the rod should not

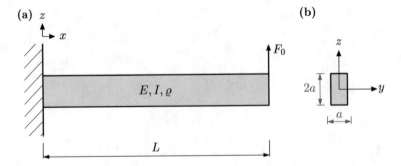

Fig. 3.11   a General configuration of the beam problem; **b** cross-sectional area

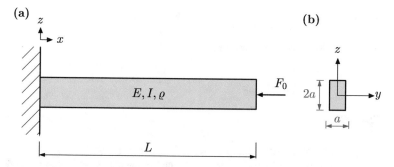

Fig. 3.12 a General configuration of the compression rod problem; **b** cross-sectional area

Fig. 3.13 a General configuration of the short beam problem; **b** cross-sectional area

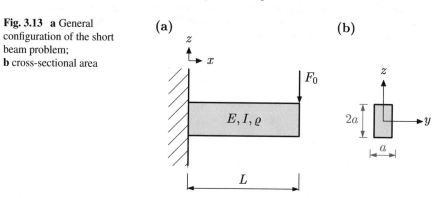

buckle in the elastic range. Some aspects of the required theories of mechanics of materials can be found in [1–3]. Use the exterior penalty function method to solve this problem.

3.8 Numerical Determination of the Optimal Design of a Short Cantilever Beam

Given is a short cantilever beam as shown in Fig. 3.13. The beam is loaded by a single force F_0 and has constant material (E, ϱ) and geometrical properties (I) along its axis. The material is isotropic and homogeneous and the beam theory for thick beams should be applied for this example, see [1–3] for details on the theory.

Given are:

- Geometrical dimension: $L = 846.33$ mm.
- Material properties of the beam: Young's modulus $E = 68948$ MPa, mass density $\varrho = 2691$ kg/m^3, initial tensile yield stress $R_{p0.2} = 247$ MPa.
- Loading: $F_0 = 2667$ N.

Determine the optimized cross-sectional dimension a under the condition that the maximum normal *and* shear stresses do not exceed the corresponding initial yield stresses. The initial shear yield stress can be approximated based on the Tresca yield condition. Consider that the normal stress has a linear distribution while the shear stress has a parabolic distribution over the beam height. Use the exterior penalty function method to solve this problem.

References

1. Boresi AP, Schmidt RJ (2003) Advanced mechanics of materials. Wiley, New York
2. Öchsner A (2014) Elasto-plasticity of frame structure elements: modeling and simulation of rods and beams. Springer, Berlin
3. Öchsner A (2016) Continuum damage and fracture mechanics. Springer, Singapore
4. Vanderplaats GN (1999) Numerical optimization techniques for engineering design. Vanderplaats Research & Development, Colorado Springs

Chapter 4
Unconstrained Functions of Several Variables

Abstract This chapter introduces two classical numerical methods to find the minimum of a unimodal function of several variables. The first method, i.e., the steepest descent method, is a typical representative of first-order methods which requires functional evaluations of the objective function and the calculation of the gradient operator. The second method, i.e., Newton's method, is a typical representative of second-order methods and requires the evaluation of the gradient operator, the Hessian matrix, as well as functional evaluations of the objective function.

4.1 General Introduction to the Unconstrained Multidimensional Optimization Problem

Let us consider in the following a scalar objective function $F(X)$, where the argument is a column matrix with n design variables: $X = \{X_1 \ X_2 \ X_3 \ \ldots \ X_n\}^{\mathrm{T}}$. The task of finding the minimum of such a function can be approached based on different methods, which are generally summarized in Table 4.1.

The gradient operator[1] is defined by the first-order derivatives

$$\nabla F(X) = \left\{ \begin{array}{c} \dfrac{\partial F(X)}{\partial X_1} \\[2mm] \dfrac{\partial F(X)}{\partial X_2} \\[1mm] \vdots \\[1mm] \dfrac{\partial F(X)}{\partial X_n} \end{array} \right\}, \tag{4.1}$$

---

[1] The gradient operator is also called the Nabla or the Hamilton operator.

A. Öchsner and R. Makvandi, *Numerical Engineering Optimization*, https://doi.org/10.1007/978-3-030-43388-8_4

Table 4.1 General classification of optimization approaches to find the minimum of an objective function. Information extracted from [5]

Approach	Characteristics
Zero-order methods	Utilize function values only
First-order methods	Utilize function values and gradient information
Second-order methods	Utilize function values, gradient information, and the Hessian matrix

and the second-order derivatives are collected in the so-called Hessian matrix [1]:

$$H(X) = \begin{bmatrix} \dfrac{\partial^2 F(X)}{\partial X_1^2} & \dfrac{\partial^2 F(X)}{\partial X_1 \partial X_2} & \cdots & \dfrac{\partial^2 F(X)}{\partial X_1 \partial X_n} \\ \dfrac{\partial^2 F(X)}{\partial X_2 \partial X_1} & \dfrac{\partial^2 F(X)}{\partial X_2^2} & \cdots & \dfrac{\partial^2 F(X)}{\partial X_2 \partial X_n} \\ \vdots & \vdots & & \vdots \\ \dfrac{\partial^2 F(X)}{\partial X_n \partial X_1} & \dfrac{\partial^2 F(X)}{\partial X_n \partial X_2} & \cdots & \dfrac{\partial^2 F(X)}{\partial X_n^2} \end{bmatrix}. \tag{4.2}$$

Let us recall that the necessary and sufficient conditions for a scalar function to have at least a local minimum can be expressed as $\frac{dF(X)}{dX} = 0$ and $\frac{d^2 F(X)}{d^2 X} > 0$. This can be generalized for the n-dimensional case to the following necessary condition using the gradient operator (4.1):

$$\nabla F(X) = \mathbf{0}. \tag{4.3}$$

The generalization of the sufficient condition means that the Hessian matrix must be positive definite, i.e. *all* eingenvalues of the matrix must be positive.

The general optimization algorithm for an unconstrained problem can be generally postulated as [5]

$$X_{\text{new}} = X_{\text{old}} + \alpha_{\text{old}}^* S_{\text{old}}, \tag{4.4}$$

where S is the search direction and α^* is a scalar multiplier which determines the amount of change for each iteration. It should be noted that it might be advantageous to normalize S by its maximum absolute component $|S_i|$.

An algorithmic realization of a general unconstrained minimization strategy for several design variables is presented in Fig. 4.1. For the one-dimensional search, the methods categorized in Table 4.1 are available.

Convergence of the algorithm as generally indicated in Fig. 4.1 ('Converged?') can be based—depending on the particular algorithm—on different criteria, see [5]:

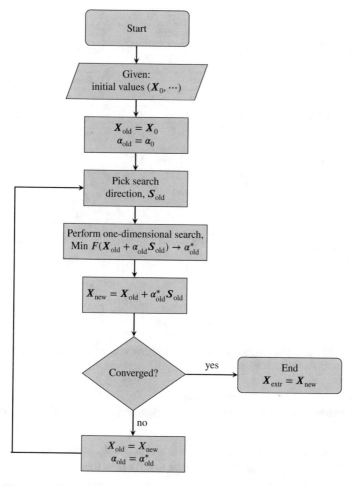

Fig. 4.1 Flowchart of a general unconstrained minimization strategy for several design variables. Adapted from [5]

- Maximum number of iterations: the iteration is terminated as soon as the iteration number exceeds a large predefined number N_{max}.
- Absolute change in objective function: the iteration is terminated as soon as the absolute difference in F is below or equal a predefined small number (e.g. $\varepsilon_{abs} = 0.001$): $|F(X_{new}) - F(X_{old})| \leq \varepsilon_{abs}$.
- Relative change in objective function: the iteration is terminated as soon as the absolute difference in F is below or equal a predefined small number (e.g. $\varepsilon_{rel} = 0.001$): $\frac{|F(X_{new}) - F(X_{old})|}{\text{Max}[|F(X_{new})|, 10^{-10}]} \leq \varepsilon_{rel}$. Alternatively, Ref. [4] indicates a value of 10^{-8} in the denominator.
- Check of the Kuhn-Tucker condition: The necessary condition for an unconstrained function reads $\nabla F(X) = 0$. The iteration is terminated as soon as each component

Table 4.2 Different convergence criteria available in the Maxima library my_funs.mac

Criteria	Declarative keyword	Comment
Maximum number of iterations	max_iter	Requires additional variable defined: maximum_iteration
Absolute changes of F	abs_change	Requires variable eps
Relative changes of F	rel_change	Requires variable eps
Kuhn-Tucker condition	Kuhn_Tucker	Requires variable eps

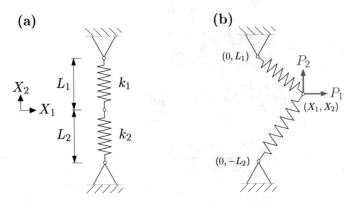

Fig. 4.2 Schematic representation of a planar two-spring system: **a** undeformed configuration; **b** deformed configuration. Adapted from [5]

of the gradient is below or equal in magnitude a predefined small number (e.g. $\varepsilon_{KT} = 0.001$).

The above mentioned convergence criteria are available in the following Maxima codes and Table 4.2 provides some information on how these criteria can be called in our Maxima modules.

Let us now have a look on a particular example with two design variables (adapted from [5]). Figure 4.2 shows a planar two-spring system where the constant spring stiffnesses are given by k_1 and k_2. In the unloaded, i.e. undeformed, configuration, the lengths of the linear springs are given by L_1 and L_2, see Fig. 4.2a. The single forces P_1 and P_2 deform the system as shown in Fig. 4.2b and the position of the load application point in the deformed configuration is designated by (X_1, X_2).

The contraction or elongation of each spring in the deformed configuration can be expressed based on Pythagoras' theorem as:

$$\Delta L_1 = \sqrt{X_1^2 + (L_1 - X_2)^2} - L_1 , \tag{4.5}$$

$$\Delta L_2 = \sqrt{X_1^2 + (L_2 + X_2)^2} - L_2 . \tag{4.6}$$

Fig. 4.3 Contour diagram of the objective function $F(X_1, X_2)$ based on the following parameters: $L_1 = 10$, $L_2 = 10$, $k_1 = 8$, $k_2 = 1$, $P_1 = 5$ and $P_2 = 5$ (consistent units assumed), adapted from [5]. Exact solution for the minimum: $X_{1,\text{extr}} = 8.631$, $X_{2,\text{extr}} = 4.533$

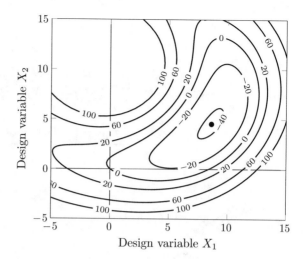

Thus, we can write the total potential energy (Π), i.e. the sum of the strain energy (Π_i) and the work done by the external loads (Π_e), as:

$$\Pi = \Pi_i + \Pi_a \,, \tag{4.7}$$

$$= \frac{1}{2} \times k_1(\Delta L_1)^2 + \frac{1}{2} \times k_2(\Delta L_2)^2 - P_1 X_1 - P_2 X_2 \,, \tag{4.8}$$

$$= \frac{1}{2} \times k_1 \left(\sqrt{X_1^2 + (L_1 - X_2)^2} - L_1 \right)^2$$

$$+ \frac{1}{2} \times k_2 \left(\sqrt{X_1^2 + (L_2 + X_2)^2} - L_2 \right)^2 \tag{4.9}$$

$$- P_1 X_1 - P_2 X_2 \,.$$

Based on the minimum total potential energy principle, the total potential energy attains a minimum for a structure in an equilibrium state [3]. Thus, the total potential energy can be taken as the objective function, i.e. $F(X_1, X_2) = \Pi(X_1, X_2)$, see Fig. 4.3 for a graphical representation.

Let us introduce in the following an additional example (adapted from [5]). Figure 4.4 shows a planar spring weight system which is composed of 6 springs (Roman numbering I to VI) and 5 different masses (Arabic numbering 2 to 6). The undeformed configuration, i.e. without the action of any of the masses, is shown in Fig. 4.4a. In this situation, all the springs are horizontally aligned and unstretched.

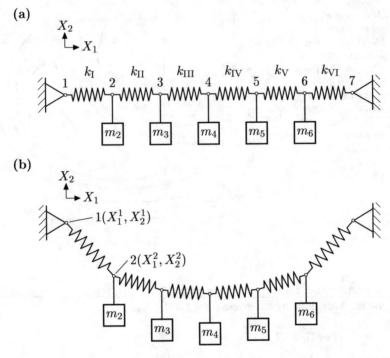

Fig. 4.4 Schematic representation of a planar spring and weight system: **a** undeformed configuration; **b** deformed configuration. Adapted from [5]

Under the action of the different masses, the systems deforms and the deformed shape is the unknown quantity. As in the previous example, the total potential energy (Π), i.e. the sum of the strain energy (Π_i) and the work done by the external loads (Π_e), can be stated as:

$$\Pi = \Pi_i + \Pi_a,$$
$$= \frac{1}{2} \times k_I (\Delta L_I)^2 + \cdots + \frac{1}{2} \times k_{VI}(\Delta L_{VI})^2 - P_2 X_2^2 - \cdots - P_6 X_2^6, \quad (4.10)$$

where the lengths of the springs can be expressed based on the deformed coordinates of the joints ($i(X_1^i, X_2^i)$ with $i = 1, \ldots, 7$) as:

$$\Delta L_I = \sqrt{\left(X_1^2 - X_1^1\right)^2 + \left(X_2^2 - X_2^1\right)^2} - L_I, \quad (4.11)$$

$$\vdots$$

$$\Delta L_{VI} = \sqrt{\left(X_1^7 - X_1^6\right)^2 + \left(X_2^7 - X_2^6\right)^2} - L_{VI}. \quad (4.12)$$

For a numerical approach, let us assume that each spring has an undeformed length of $L_i = 10\,\mathrm{m}$ ($i = \mathrm{I}, \cdots, \mathrm{VI}$) and that the stiffnesses are given as:

$$k_{\mathrm{I}} = 500 + 200 \left(\frac{5}{3} - 1\right)^2 \; \mathrm{N/m}, \qquad (4.13)$$

$$k_{\mathrm{II}} = 500 + 200 \left(\frac{5}{3} - 2\right)^2 \; \mathrm{N/m}, \qquad (4.14)$$

$$\vdots$$

$$k_{\mathrm{VI}} = 500 + 200 \left(\frac{5}{3} - 6\right)^2 \; \mathrm{N/m}. \qquad (4.15)$$

The vertical forces, which are acting at the joints of the springs, follow the following convention:

$$P_2 = m_2 g = -50 \times 1 \; \mathrm{N}, \qquad (4.16)$$
$$P_3 = m_2 g = -50 \times 2 \; \mathrm{N}, \qquad (4.17)$$

$$\vdots$$

$$P_6 = m_6 g = -50 \times 5 \; \mathrm{N}. \qquad (4.18)$$

Based on the minimum total potential energy principle, the total potential energy attains a minimum for a structure in an equilibrium state. Thus, the total potential energy can be taken as the objective function, i.e. $F(X_1^1, X_2^1, \cdots, X_1^7, X_2^7) = \Pi(X_1^1, X_2^1, \cdots, X_1^7, X_2^7)$. Since the first and last joints do not move, we can simplify the objective functions to

$$F(X_1^2, X_2^2, \cdots, X_1^6, X_2^6), \qquad (4.19)$$

which means that we have 10 design variables $(X_1^2, \cdots, X_2^6)$.

The above described spring problems will be numerically solved in the following sections based on different methods.

4.2 First-Order Methods

Let us introduce here a classical first-order method, i.e. the steepest descent method, see [5]. In this method, the search direction is taken as the negative gradient of the objective function, i.e.

$$S = -\nabla F(X),\tag{4.20}$$

and the update rule can be written based on Eq. (4.4) as:

$$X_{new} = X_{old} - \alpha^*_{old}\nabla F(X_{old}).\tag{4.21}$$

Following the flowchart given in Fig. 4.1, an objective function of several design variables can be minimized. It should be noted here that literature suggests more efficient first-order methods such as the conjugate direction method or variable metric methods (see [5] for details). Nevertheless, the steepest descent method is a fundamental approach, which serves in many other methods as starting point.

4.1 Numerical Determination of the Equilibrium Position of a Planar Two-Spring System Based on the Steepest Descent Method

Determine the equilibrium position of the two-spring system shown in Figs. 4.2 and 4.3 based on the steepest descent method. Start the iteration from three different initial points, i.e. $X_0 = \begin{bmatrix} -4 & 4 \end{bmatrix}^T$, $X_0 = \begin{bmatrix} 1 & 1 \end{bmatrix}^T$, and $X_0 = \begin{bmatrix} 2.5 & 12.5 \end{bmatrix}^T$, and use different initial scalar multipliers, i.e. $\alpha_0 = 1.0$ and $\alpha_0 = 2.0$. Stop the iteration as soon as the Kuhn-Tucker condition is fulfilled based on $\varepsilon_{KT} = 0.001$.

4.1 Solution

The following Listing 4.1 shows the entire wxMaxima code for the determination of the minimum of the objective function given in Eq. (4.9) based on the steepest descend method.

The graphical representation of the iteration history is given in Figs. 4.5 and 4.6. Comparing both figures and the convergence summary in Table 4.3, it can be concluded that the initial starting multiplier has for the given parameters no influence on the required number of iterations for convergence. Nevertheless, independent of the initial points and the initial starting multipliers, the minimum is correctly obtained for all cases.

```
(% i19)   load("my_funs.mac")$

          fpprintprec:6$
          ratprint: false$

          L[1] : 10$
          L[2] : 10$
          P[1] : 5$
          P[2] : 5$
          k[1] : 8$
          k[2] : 1$

          eps : 1/1000$

          deltaL[1](X) := sqrt(X[1]^2+(L[1]-X[2])^2)-L[1]$
          deltaL[2](X) := sqrt(X[1]^2+(L[2]+X[2])^2)-L[2]$

          f[1](X) := 0.5*k[1]*deltaL[1](X)^2 - P[1]*X[1]$
          f[2](X) := 0.5*k[2]*deltaL[2](X)^2 - P[2]*X[2]$
          func(X) := f[1](X) + f[2](X)$

          no_of_vars : 2$

          X_0s : [[X[1]=-4,X[2]=4],[X[1]=1,X[2]=1],[X[1]=2.5,X[2]=12.5] ]$
          alpha_0 : 1$

          for i : 1 thru length(X_0s) do (
             print("==============================="),
             print(" "),
             print("For",X["0"] = X_0s[i]),
             print(" "),
             X_new : steepest_multi_variable_unconstrained(func,no_of_vars,X_0s[i],
              alpha_0,eps,"Kuhn_Tucker",true),
             print(X["1"] = rhs(X_new[1])),
             print(X["2"] = rhs(X_new[2])),
             print(" ")
          )$
```

===============================

For X[0]=[X[1]=-4,X[2]=4]

i=1 X=[X[1]=-4.74356,X[2]=1.82043] func(X)=19.5463
i=2 X=[X[1]=2.17286,X[2]=-0.539103] func(X)=-5.81089
i=3 X=[X[1]=7.27634,X[2]=14.4205] func(X)=20.1884
i=4 X=[X[1]=9.70551,X[2]=13.5918] func(X)=4.28329
i=5 X=[X[1]=5.98774,X[2]=2.69405] func(X)=-34.0399
i=6 X=[X[1]=6.96237,X[2]=2.36155] func(X)=-37.4025
i=7 X=[X[1]=7.42977,X[2]=3.73166] func(X)=-39.7435
i=8 X=[X[1]=7.97669,X[2]=3.54509] func(X)=-40.9806
i=9 X=[X[1]=8.23398,X[2]=4.29928] func(X)=-41.5369
i=10 X=[X[1]=8.44552,X[2]=4.22711] func(X)=-41.7329

```
i=11 X=[X[1]=8.53081,X[2]=4.47713] func(X)=-41.7894
i=12 X=[X[1]=8.58786,X[2]=4.45767] func(X)=-41.8038
i=13 X=[X[1]=8.60903,X[2]=4.51973] func(X)=-41.8072
i=14 X=[X[1]=8.62221,X[2]=4.51523] func(X)=-41.808
i=15 X=[X[1]=8.62698,X[2]=4.52923] func(X)=-41.8082
i=16 X=[X[1]=8.6299,X[2]=4.52824] func(X)=-41.8082
i=17 X=[X[1]=8.63095,X[2]=4.53132] func(X)=-41.8082
i=18 X=[X[1]=8.63159,X[2]=4.5311] func(X)=-41.8082
i=19 X=[X[1]=8.63182,X[2]=4.53178] func(X)=-41.8082
i=20 X=[X[1]=8.63196,X[2]=4.53173] func(X)=-41.8082
Converged after 20 iterations!
X[1]=8.63196
X[2]=4.53173v


==============================


For X[0]=[X[1]=1,X[2]=1]

i=1 X=[X[1]=2.57176,X[2]=0.0274139] func(X)=-12.5767
i=2 X=[X[1]=3.67502,X[2]=1.81036] func(X)=-20.4301
i=3 X=[X[1]=4.96798,X[2]=1.01029] func(X)=-27.4358
i=4 X=[X[1]=6.98568,X[2]=4.27103] func(X)=-35.2137
i=5 X=[X[1]=8.00284,X[2]=3.64163] func(X)=-41.1147
i=6 X=[X[1]=8.57244,X[2]=4.56216] func(X)=-41.7896
i=7 X=[X[1]=8.62904,X[2]=4.52714] func(X)=-41.8082
i=8 X=[X[1]=8.63201,X[2]=4.53194] func(X)=-41.8082
Converged after 8 iterations!
X[1]=8.63201
X[2]=4.53194


==============================


For X[0]=[X[1]=2.5,X[2]=12.5]

i=1 X=[X[1]=7.14027,X[2]=15.8506] func(X)=28.8429
i=2 X=[X[1]=11.0407,X[2]=10.4488] func(X)=-15.4037
i=3 X=[X[1]=3.03907,X[2]=4.67111] func(X)=33.6282
i=4 X=[X[1]=5.32357,X[2]=1.50727] func(X)=-30.5635
i=5 X=[X[1]=8.48222,X[2]=3.78802] unc(X)=-41.149
i=6 X=[X[1]=8.28049,X[2]=4.06739] func(X)=-41.6069
i=7 X=[X[1]=8.57535,X[2]=4.28029] func(X)=-41.7398
i=8 X=[X[1]=8.51214,X[2]=4.36783] func(X)=-41.7836
i=9 X=[X[1]=8.61055,X[2]=4.43888] func(X)=-41.7991
i=10 X=[X[1]=8.58771,X[2]=4.4705] func(X)=-41.8048
i=11 X=[X[1]=8.62383,X[2]=4.49658] func(X)=-41.8069
i=12 X=[X[1]=8.61523,X[2]=4.50849] func(X)=-41.8077
i=13 X=[X[1]=8.6289,X[2]=4.51836] func(X)=-41.808
i=14 X=[X[1]=8.62561,X[2]=4.52292] func(X)=-41.8082
i=15 X=[X[1]=8.63085,X[2]=4.5267] func(X)=-41.8082
i=16 X=[X[1]=8.62958,X[2]=4.52845] func(X)=-41.8082
i=17 X=[X[1]=8.6316,X[2]=4.5299] func(X)=-41.8082
```

```
i=18 X=[X[1]=8.63111,X[2]=4.53057] func(X)=-41.8082
i=19 X=[X[1]=8.63188,X[2]=4.53113] func(X)=-41.8082
i=20 X=[X[1]=8.6317,X[2]=4.53139] func(X)=-41.8082
Converged after 20 iterations!
X[1]=8.6317
X[2]=4.53139
```

Module 4.1: Numerical determination of the minimum for the function (4.9) based on the steepest descent method

Fig. 4.5 Iteration history in the contour diagram of the objective function $F(X_1, X_2)$ according to Eq. (4.9) based on the steepest descent method for $\alpha_0 = 1.0$: **a** $X_0 = [1\ 1]^T$ and **b** $X_0 = [2.5\ 12.5]^T$

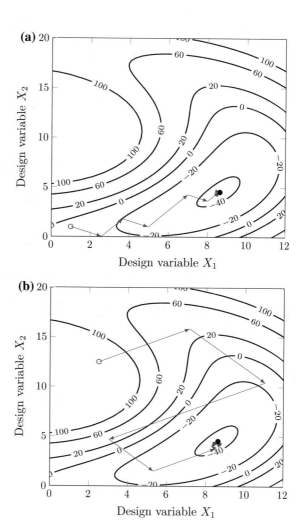

Fig. 4.6 Iteration history in
the contour diagram of the
objective function
$F(X_1, X_2)$ according to
Eq. (4.9) based on the
steepest descent method for
$\alpha_0 = 2.0$: **a** $X_0 = [1\ 1]^T$ and
b $X_0 = [2.5\ 12.5]^T$

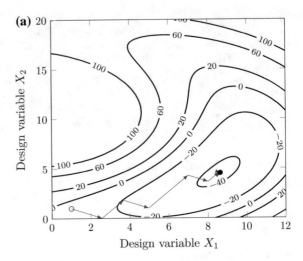

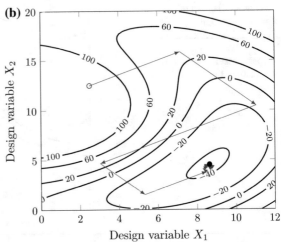

Table 4.3 Convergence history for the function (4.9) based on the steepest descent method

Initial points	Iterations for convergence	
	$\alpha_0 = 1.0$	$\alpha_0 = 2.0$
$X_0 = \begin{bmatrix} -4\ 4 \end{bmatrix}^T$	20	20
$X_0 = \begin{bmatrix} 1\ 1 \end{bmatrix}^T$	8	8
$X_0 = \begin{bmatrix} 2.5\ 12.5 \end{bmatrix}^T$	20	20

4.2 Numerical Determination of the Equilibrium Position of a Planar Spring Weight System Based on the Steepest Descent Method

Determine the equilibrium position of the planar spring weight system shown in Fig. 4.4 based on the steepest descent method. Start the iteration from the undeformed configuration and stop the iteration as soon as the Kuhn-Tucker condition is fulfilled based on $\varepsilon_{KT} = 0.001$.

4.2 Solution

The following Listing 4.2 shows the entire wxMaxima code for the determination of the minimum of the objective function given in Eq. (4.10) based on the steepest descent method.

```
(% i56)   load("my_funs.mac")$

          fpprintprec:8$
          ratprint: false$

          X1 : 0$
          X2 : 0$
          X13 : 60$
          X14 : 0$

          L[1] : 10$
          L[2] : 10$
          L[3] : 10$
          L[4] : 10$
          L[5] : 10$
          L[6] : 10$

          deltaL[1](X) := sqrt( (X[1]-X1)^2 + (X[2]-X2)^2 ) - L[1]$
          deltaL[2](X) := sqrt( (X[3]-X[1])^2 + (X[4]-X[2])^2 ) - L[2]$
          deltaL[3](X) := sqrt( (X[5]-X[3])^2 + (X[6]-X[4])^2 ) - L[3]$
          deltaL[4](X) := sqrt( (X[7]-X[5])^2 + (X[8]-X[6])^2 ) - L[4]$
          deltaL[5](X) := sqrt( (X[9]-X[7])^2 + (X[10]-X[8])^2 ) - L[5]$
          deltaL[6](X) := sqrt( (X13-X[9])^2 + (X14-X[10])^2 ) - L[6]$

          k[1] : 500 + 200*((5/3)-1)^2$
          k[2] : 500 + 200*((5/3)-2)^2$
          k[3] : 500 + 200*((5/3)-3)^22$
          k[4] : 500 + 200*((5/3)-4)^2$
          k[5] : 500 + 200*((5/3)-5)^2$
          k[6] : 500 + 200*((5/3)-6)^2$
```

```
P[2] : -50 * 1$
P[3] : -50 * 2$
P[4] : -50 * 3$
P[5] : -50 * 4$
P[6] : -50 * 5$

f[1](X) := 0.5*k[1]*deltaL[1](X)^2$
f[2](X) := 0.5*k[2]*deltaL[2](X)^2 - P[2]*X[2]$
f[3](X) := 0.5*k[3]*deltaL[3](X)^2 - P[3]*X[4]$
f[4](X) := 0.5*k[4]*deltaL[4](X)^2 - P[4]*X[6]$
f[5](X) := 0.5*k[5]*deltaL[5](X)^2 - P[5]*X[8]$
f[6](X) := 0.5*k[6]*deltaL[6](X)^2 - P[6]*X[10]$

func(X) := f[1](X) + f[2](X) + f[3](X) + f[4](X) + f[5](X) + f[6](X)$

eps : 1/1000$

no_of_vars : 10$

X_0 : [X[1]=10,X[2]=0,X[3]=20,X[4]=0,X[5]=30,X[6]=0,X[7]=40,X[8]=0,X[9]=50,
                    X[10]=0]$
alpha_0 : 3$

X_new : steepest_multi_variable_unconstrained(func,no_of_vars,X_0,alpha_0,
                    eps,"Kuhn_Tucker",true)$

print(X["1"]^"1" = X1)$
print(X["2"]^"1" = X2)$
print(X["1"]^"2" = rhs(X_new[1]))$
print(X["2"]^"2" = rhs(X_new[2]))$
print(X["1"]^"3" = rhs(X_new[3]))$
print(X["2"]^"3" = rhs(X_new[4]))$
print(X["1"]"4" = rhs(X_new[5]))$
print(X["2"]^"4" = rhs(X_new[6]))$
print(X["1"]^"5" = rhs(X_new[7]))$
print(X["2"]^"5" = rhs(X_new[8]))$
print(X["1"]^"6" = rhs(X_new[9]))$
print(X["2"]^"6" = rhs(X_new[10]))$
print(X["1"]^"7" = X13)$
print(X["2"]^"7" = X14)$

i=1
X=[X[1]=10.0,X[2]=-0.60395329,X[3]=20.0,X[4]=-1.2079066,X[5]=30.0,X[6]=-
1.8118599,X[7]=40.0,X[8]=-2.4158132,X[9]=50.0,X[10]=-3.0197665] func(X)=-1236.5737
i=2
X=[X[1]=9.9921159,X[2]=-0.928581,X[3]=20.03942,X[4]=-1.8604951,X[5]=30.086725,
X[6]=-2.7924093,X[7]=40.134029,X[8]=-3.7243234,X[9]=61.492054,X[10]=-1.0582941]
func(X)=321156.12
```

```
i=3
X=[X[1]=9.9991511,X[2]=-0.94255643,X[3]=20.047419,X[4]=-1.8878848,X[5]=30.104321,
X[6]=-2.834013,X[7]=48.391004,X[8]=-2.7386415,X[9]=60.754588,X[10]=-7.520883]
func(X)=78589.369

...

i=1318
X=[X[1]=10.355028,X[2]=-4.280037,X[3]=21.086663,X[4]=-7.8973358,X[5]=31.686934,
X[6]=-9.8535827,X[7]=42.089803,X[8]=-9.393391,X[9]=51.771341,X[10]=-6.0117915]
func(X)=-4416.3842
i=1319
X=[X[1]=10.355028,X[2]=-4.2800371,X[3]=21.086663,X[4]=-7.897336,X[5]=31.686934,
X[6]=-9.8535829,X[7]=42.089803,X[8]=-9.3933911,X[9]=51.77134,X[10]=-6.0117917]
func(X)=-4416.3842
Converged after 1319 iterations!
X[1]^1=0
X[2]^1=0
X[1]^2=10.355028
X[2]^2=-4.2800371
X[1]^3=21.086663
X[2]^3=-7.897336
X[1]^4=31.686934
X[2]^4=-9.8535829
X[1]^5=42.089803
X[2]^5=-9.3933911
X[1]^6=51.77134
X[2]^6=-6.0117917
X[1]^7=60
X[2]^7=0
```

Module 4.2: Numerical determination of the minimum for the function (4.10) based on the steepest descent method

4.3 Second-Order Methods

Let us introduce here a classical second-order method, i.e. Newton's method, see [5]. Following the approach for functions of single arguments in Sect. 2.3, let us write a second-order Taylor series expansion [2] of the objective function about X_0, i.e.

$$F(X) \approx F(X_0) + \nabla F(X)|_{X_0} (X - X_0) + \frac{1}{2}(X - X_0)^{\mathrm{T}} H(X)|_{X_0} (X - X_0).$$
(4.22)

where ∇ is the gradient operator (see Eq. 4.1) and H is the Hessian matrix (see Eq. (4.2)).

Differentiating Eq. (4.22) with respect to X and neglecting differences of higher order gives finally the following expression under consideration of the necessary condition for a local minimum:

$$\nabla F(X) = \nabla F(X)|_{X_0} + H(X)|_{X_0}\,(X - X_0) \overset{!}{=} 0 \,. \tag{4.23}$$

The last expression can be rearranged to give the following iteration scheme:

$$X = X_0 - \left(H(X)|_{X_0}\right)^{-1} \nabla F(X)|_{X_0} \,. \tag{4.24}$$

A comparison with the general approach given in Eq. (4.4) allows the identification of the search direction and the scalar multiplier as:

$$S = -\left(H(X)\right)^{-1} \nabla F(X) \,, \tag{4.25}$$

$$\alpha^* = 1 \,. \tag{4.26}$$

4.3 Numerical Determination of the Equilibrium Position of a Planar Two-Spring System Based on Newton's Method

Determine the equilibrium position of the two-spring system shown in Figs. 4.2 and 4.3 based on Newton's method. Start the iteration from different initial points, i.e. $X_0 = [-4\ 4]^T$, $X_0 = [1\ 1]^T$ and $X_0 = [2.5\ 12.5]^T$, and use different initial scalar multipliers, i.e. $\alpha_0 = 1.0$ and $\alpha_0 = 2.0$. Stop the iteration as soon as the Kuhn-Tucker condition is fulfilled based on $\varepsilon_{KT} = 0.001$.

4.3 Solution

The following Listing 4.3 shows the entire wxMaxima code for the determination of the minimum of the objective function given in Eq. (4.9) based on Newtons' method. The graphical representation of the iteration history is given in Figs. 4.7 and 4.8.

Analyzing the particular data, it turned out that the combination of $\alpha_0 = 1.0$ and $X_0 = [1\ 1]^T$ leads to no convergence in minimizing the α value, i.e. exceeding a maximum number of 50 iterations. In this case, our algorithm updates the value of X based on the initial value of the scalar multiplier. For all other cases, we obtain convergence for each iteration step and the initial value of the scalar multiplier has no influence on the iteration path, see Figs. 4.7 and 4.8 and Table 4.4.

```
(% i21)    load("my_funs.mac")$

           fpprintprec:6$
           ratprint: false$

           L[1] : 10$
           L[2] : 10$
           P[1] : 5$
           P[2] : 5$
           k[1] : 8$
           k[2] : 1$

           eps : 1/1000$

           deltaL[1](X) := sqrt(X[1]^2+(L[1]-X[2])^2)-L[1]$
           deltaL[2](X) := sqrt(X[1]^2+(L[2]+X[2])^2)-L[2]$

           f[1](X) := 0.5*k[1]*deltaL[1](X)^2 - P[1]*X[1]$
           f[2](X) := 0.5*k[2]*deltaL[2](X)^2 - P[2]*X[2]$
           func(X) := f[1](X) + f[2](X)$

           no_of_vars : 2$

           X_0 : [X[1]=1,X[2]=1]$
           alpha_0 : 2$

           X_new : Newton_multi_variable_unconstrained(func,no_of_vars,X_0,alpha_0,
                               eps,"Kuhn_Tucker",true)$
           print(X["1"] = rhs(X_new[1]))$
           print(X["2"] = rhs(X_new[2]))$

i=1 X=[X[1]=7.10199,X[2]=1.93307] func(X)=-35.3863
i=2 X=[X[1]=8.76715,X[2]=4.41209] func(X)=-41.674
i=3 X=[X[1]=8.63068,X[2]=4.52983] func(X)=-41.8082
i=4 X=[X[1]=8.63207,X[2]=4.53191] func(X)=-41.8082
Converged after 4 iterations!
X[1]=8.63207
X[2]=4.53191
```

Module 4.3: Numerical determination of the minimum for the function (4.9) based on the Newton's method

4.4 Numerical Determination of the Equilibrium Position of a Planar Spring Weight System Based on Newton's Method

Determine the equilibrium position of the planar spring weight system shown in Fig. 4.4 based on Newton's method. Start the iteration from the undeformed

Fig. 4.7 Iteration history in the contour diagram of the objective function $F(X_1, X_2)$ according to Eq. (4.9) based on Newton's method for $\alpha_0 = 1.0$: **a** $X_0 = [1\ 1]^T$ and **b** $X_0 = [2.5\ 12.5]^T$

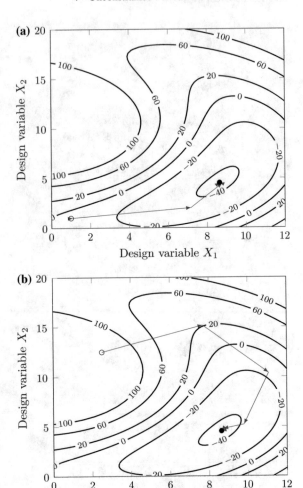

Table 4.4 Convergence history for the function (4.9) based on Newton's method

Initial points	Iterations for convergence		Comment
	$\alpha_0 = 1.0$	$\alpha_0 = 2.0$	
$X_0 = \begin{bmatrix} -4\ 4 \end{bmatrix}^T$	7	7	Identical paths
$X_0 = \begin{bmatrix} 1\ 1 \end{bmatrix}^T$	5	4	Convergence failure for $\alpha_0 = 1.0$ and $i = 1$
$X_0 = \begin{bmatrix} 2.5\ 12.5 \end{bmatrix}^T$	6	6	Identical paths

Fig. 4.8 Iteration history in the contour diagram of the objective function $F(X_1, X_2)$ according to Eq. (4.9) based on Newton's method for $\alpha_0 = 2.0$: **a** $X_0 = [1\ 1]^T$ and **b** $X_0 = [2.5\ 12.5]^T$

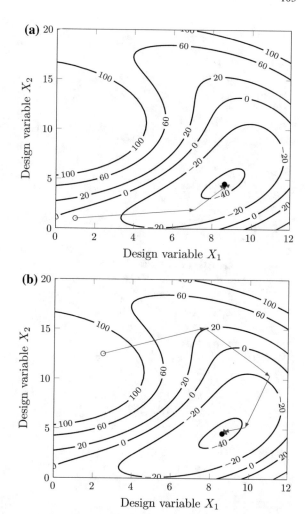

configuration and stop the iteration as soon as the Kuhn-Tucker condition is fulfilled based on $\varepsilon_{KT} = 0.001$.

4.4 Solution

The following Listing 4.4 shows the entire wxMaxima code for the determination of the minimum of the objective function given in Eq. (4.10) based on Newton's method.

Based on the numerical values obtained in Listing 4.4, the deformed shape is illustrated in Fig. 4.9.

```
(% i56)    load("my_funs.mac")$

           fpprintprec:8$
           ratprint: false$

           X1 : 0$
           X2 : 0$
           X13 : 60$
           X14 : 0$

           L[1] : 10$
           L[2] : 10$
           L[3] : 10$
           L[4] : 10$
           L[5] : 10$
           L[6] : 10$

           deltaL[1](X) := sqrt( (X[1]-X1)^2 + (X[2]-X2)^2 ) - L[1]$
           deltaL[2](X) := sqrt( (X[3]-X[1])^2 + (X[4]-X[2])^2 ) - L[2]$
           deltaL[3](X) := sqrt( (X[5]-X[3])^2 + (X[6]-X[4])^2 ) - L[3]$
           deltaL[4](X) := sqrt( (X[7]-X[5])^2 + (X[8]-X[6])^2 ) - L[4]$
           deltaL[5](X) := sqrt( (X[9]-X[7])^2 + (X[10]-X[8])^2 ) - L[5]$
           deltaL[6](X) := sqrt( (X13-X[9])^2 + (X14-X[10])^2 ) - L[6]$

           k[1] : 500 + 200*((5/3)-1)^2$
           k[2] : 500 + 200*((5/3)-2)^2$
           k[3] : 500 + 200*((5/3)-3)^22$
           k[4] : 500 + 200*((5/3)-4)^2$
           k[5] : 500 + 200*((5/3)-5)^2$
           k[6] : 500 + 200*((5/3)-6)^2$

           P[2] : -50 * 1$
           P[3] : -50 * 2$
           P[4] : -50 * 3$
           P[5] : -50 * 4$
           P[6] : -50 * 5$

           f[1](X) := 0.5*k[1]*deltaL[1](X)^2$
           f[2](X) := 0.5*k[2]*deltaL[2](X)^2 - P[2]*X[2]$
           f[3](X) := 0.5*k[3]*deltaL[3](X)^2 - P[3]*X[4]$
           f[4](X) := 0.5*k[4]*deltaL[4](X)^2 - P[4]*X[6]$
           f[5](X) := 0.5*k[5]*deltaL[5](X)^2 - P[5]*X[8]$
           f[6](X) := 0.5*k[6]*deltaL[6](X)^2 - P[6]*X[10]$

           func(X) := f[1](X) + f[2](X) + f[3](X) + f[4](X) + f[5](X) + f[6](X)$

           eps : 1/1000$

           no_of_vars : 10$
```

```
X_0 : [X[1]=10,X[2]=0,X[3]=20,X[4]=0,X[5]=30,X[6]=0,X[7]=40,X[8]=0,X[9]=50,
                X[10]=0]$
alpha_0 : 2$

X_new : Newton_multi_variable_unconstrained(func,no_of_vars,X_0,alpha_0,
                eps,"Kuhn_Tucker",true)$
print(X["1"]^"1" = X1)$
print(X["2"]^"1" = X2)$
print(X["1"]^"2" = rhs(X_new[1]))$
print(X["2"]^"2" = rhs(X_new[2]))$
print(X["1"]^"3" = rhs(X_new[3]))$
print(X["2"]^"3" = rhs(X_new[4]))$
print(X["1"]^"4" = rhs(X_new[5]))$
print(X["2"]^"4" = rhs(X_new[6]))$
print(X["1"]^"5" = rhs(X_new[7]))$
print(X["2"]^"5" = rhs(X_new[8]))$
print(X["1"]^"6" = rhs(X_new[9]))$
print(X["2"]^"6" = rhs(X_new[10]))$
print(X["1"]"7" = X13)$
print(X["2"]"7" = X14)$
```

```
i=1
X=[X[1]=10.0,X[2]=-1.2871389,X[3]=20.0,X[4]=-2.5742778,X[5]=30.0,X[6]=-
3.8614167,X[7]=40.0,X[8]=-5.1485556,X[9]=50.0,X[10]=-6.4356946]
func(X)=4097.9519
i=2
X=[X[1]=9.5096927,X[2]=-5.9789985,X[3]=19.264306,X[4]=-
10.165087,X[5]=29.480655,X[6]=-10.300575,X[7]=39.890377,X[8]=-
8.6078911,X[9]=50.356745,X[10]=-6.3192663] func(X)=311.7195
i=3
X=[X[1]=10.080803,X[2]=-4.6020755,X[3]=20.932813,X[4]=-
7.3011942,X[5]=31.490618,X[6]=-10.469822,X[7]=41.717567,X[8]=-
8.8445816,X[9]=51.551006,X[10]=-5.8383113] func(X)"=-4237.7226
i=4
X=[X[1]=10.364016,X[2]=-4.2113406,X[3]=21.099806,X[4]=-
7.827711,X[5]=31.680025,X[6]=-10.014621,X[7]=42.135829,X[8]=-
9.529127,X[9]=51.791522,X[10]=-6.0470651] func(X)=-4411.3249 i=5
X=[X[1]=10.354884,X[2]=-4.2818091,X[3]=21.086618,X[4]=-
7.898806,X[5]=31.688253,X[6]=-9.8577926,X[7]=42.090135,X[8]=-
9.3953089,X[9]=51.771401,X[10]=-6.0118882] func(X)=-4416.3813
i=6
X=[X[1]=10.355026,X[2]=-4.2800515,X[3]=21.086663,X[4]=-
7.8973568,X[5]=31.686936,X[6]=-9.8536027,X[7]=42.089806,X[8]=-
9.3934028,X[9]=51.77134,X[10]=-6.0117922] func(X)=-4416.3842
```

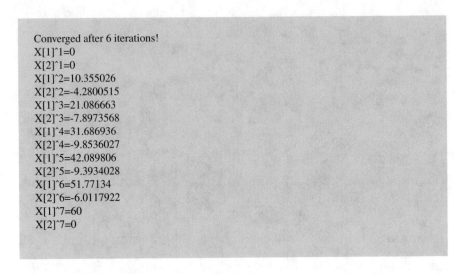

```
Converged after 6 iterations!
X[1]^1=0
X[2]^1=0
X[1]^2=10.355026
X[2]^2=-4.2800515
X[1]^3=21.086663
X[2]^3=-7.8973568
X[1]^4=31.686936
X[2]^4=-9.8536027
X[1]^5=42.089806
X[2]^5=-9.3934028
X[1]^6=51.77134
X[2]^6=-6.0117922
X[1]^7=60
X[2]^7=0
```

Module 4.4: Numerical determination of the minimum for the function (4.10) based on the Newton's method

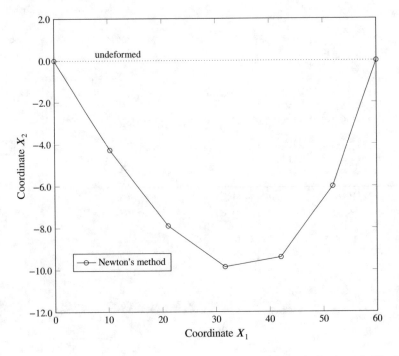

Fig. 4.9 Deformed shape of the planar spring weight system as schematically shown in Fig. 4.4

Fig. 4.10 Contour map of
the ellipsoid given in
Eq. (4.27) with $a = b = 1$
and $c = 1.5$. Exact solution
for the minimum:
$X_{1,\mathrm{extr}} = 0.0$, $X_{2,\mathrm{extr}} = 0.0$

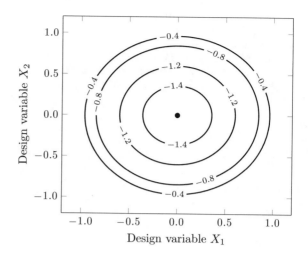

4.4 Supplementary Problems

4.5 Numerical Determination of the Minimum of an Unconstrained Function with Two Variables Based on Newton's Method

Determine based on Newton's method the minimum of the ellipsoid represented by the following equation

$$F(X_1, X_1) = -\sqrt{c^2 \left(1 - \frac{X_1^2}{a^2} - \frac{X_2^2}{b^2}\right)}, \qquad (4.27)$$

where the parameters are given by $a = b = 1$ and $c = 1.5$ (see Fig. 4.10 for a graphical representation). Start the iteration from different initial points, i.e. $X_0 = \begin{bmatrix} -0.5 & 0.5 \end{bmatrix}^T$ and $X_0 = \begin{bmatrix} 0.0 & 0.75 \end{bmatrix}^T$, and use as scalar multipliers $\alpha_0 = 1.0$. Stop the iteration as soon as the Kuhn-Tucker condition is fulfilled based on $\varepsilon_{\mathrm{KT}} = 0.001$.

References

1. Jeffrey A, Dai H-H (2008) Handbook of mathematical formulas and integrals. Academic Press, Burlington
2. Öchsner A (2014) Elasto-plasticity of frame structure elements: modeling and simulation of rods and beams. Springer, Berlin

3. Öchsner A, Merkel M (2018) One-dimensional finite elements: an introduction to the FE method. Springer, Cham
4. Schumacher A (2013) Optimierung mechanischer Strukturen: grundlagen und industrielle Anwendungen. Springer Vieweg, Berlin
5. Vanderplaats GN (1999) Numerical optimization techniques for engineering design. Vanderplaats Research & Development, Colorado Springs

Chapter 5
Constrained Functions of Several Variables

Abstract This chapter introduces a classical method for the numerical determination of the minimum of unimodal functions of several variable. The exterior penalty function method is described as a typical and efficient method to numerically solve such problems. Based on a pseudo-objective function, the problem can be treated as an unconstrained problem as covered in the previous chapter.

5.1 General Introduction to the Constrained Multidimensional Optimization Problem

The general problem statement of a constrained n-dimensional optimization procedure [10] is to minimize the objective function

$$F(X), \tag{5.1}$$

subject to the following constraints

$$
\begin{array}{lll}
g_j(X) \leq 0 & j = 1, m & \text{inequality constraints,} \\
h_k(X) = 0 & k = 1, l & \text{equality constraints,}
\end{array}
\tag{5.2}
$$

where the column matrix of the n design variables is given by:

$$
X = \begin{Bmatrix} X_1 \\ X_2 \\ X_3 \\ \vdots \\ X_n \end{Bmatrix}.
\tag{5.3}
$$

A. Öchsner and R. Makvandi, *Numerical Engineering Optimization*, https://doi.org/10.1007/978-3-030-43388-8_5

Fig. 5.1 Algorithm for a
constrained function of n
variables based on the
exterior penalty function
method, adapted from [10]

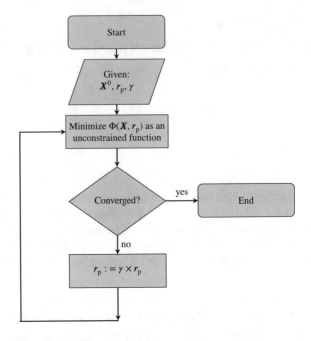

In case that side constraints, i.e. $X_i^{min} \leq X_i \leq X_i^{max}$ $(i = 1, n)$, should be considered
to limit the design space, the unconstrained minimization algorithm must be modified
by setting any design variable X_i to its lower or upper bound if Eq. (4.4) gives a value
outside the bound during the one-dimensional search.

The solution procedures will follow the ideas introduced in Chap. 3 for constrained
functions of one variable, i.e. to solve the constrained n-dimensional problem as an
unconstrained problem but to provide some penalty to limit the constraint violations.
As in the one-dimensional case, only a moderate penalty is applied at the initial stage
and the penalty increases as the optimization proceeds. Thus, several unconstrained
optimization problems will be solved. The expressions *sequential unconstrained minimization techniques* or *SUMIT* are known in literature to describe such approaches,
see [10] for details. In generalization of Fig. 3.3 for constrained functions of one
variable, we can refer to Fig. 5.1 as the general approach.

The common approach in solving such a constrained n-dimensional design problem is to formulate again a so-called pseudo-objective function

$$\Phi(X, r_p) = F(X) + r_p \times P(X), \tag{5.4}$$

where $F(X)$ is the original objective function, r_p is the scalar (non-negative) penalty
parameter or penalty multiplier and $P(X)$ is the penalty function. The pseudo-
objective function can now be treated as an unconstrained function based on the
methods presented in Chap. 4.

5.2 The Exterior Penalty Function Method

In generalization of Eq. (3.8), we can state the penalty function for the n-dimensional case as:

$$P(X) = \sum_{j=1}^{m} \left\{ \max \left[0, g_j(X) \right] \right\}^2 + \sum_{k=1}^{l} [h_k(X)]^2 \tag{5.5}$$

$$= \sum_{j=1}^{m} \delta_j \left\{ g_j(X) \right\}^2 + \sum_{k=1}^{l} [h_k(X)]^2, \tag{5.6}$$

where again no penalty is applied as long as all constraints are fulfilled. Squaring the penalty ensures that the pseudo-objective function has a continuous slope at the constraint boundary.

5.1 Numerical Determination of the Minimum of a Function with Two Variables Under Consideration of Two Constraints

Given is the objective function[1]

$$F(X) = X_1 + X_2, \tag{5.7}$$

which is constrained by the following two inequality conditions:

$$g_1(X) = -2 + X_1 - 2X_2 \le 0, \tag{5.8}$$

$$g_2(X) = 8 - 6X_1 + X_1^2 - X_2 \le 0. \tag{5.9}$$

Use the exterior penalty function method to determine the minimum. Evaluate the approaches based on the steepest descent method for the one-dimensional search. A graphical representation of the considered design space is given in Fig. 5.2. It can be easily concluded from this figure that the minimum is given by the left intersection point of g_1 and g_2, i.e., $X_{1,\text{extr}} = 2.0$, $X_{2,\text{extr}} = 0.0$. Nevertheless, the *numerical* solution approach should be the focus of the following evaluations.

5.1 Solution

The pseudo-objective function can be written for this problem as

$$\Phi(X, r_p) = X_1 + X_2 + r_p \left\{ \left[\max \left(0, -2 + X_1 - 2X_2 \right) \right]^2 \right.$$
$$\left. + \left[\max \left(0, 8 - 6X_1 + X_1^2 - X_2 \right) \right]^2 \right\}, \tag{5.10}$$

or more explicitly for the different ranges:

---

[1]This example is adapted from [10].

Fig. 5.2 Contour diagram of the objective function $F(X_1, X_2)$ and the limitations due to two inequality constraints (see Eqs. 5.7–5.9), adapted from [10]. Exact solution for the minimum: $X_{1,\text{extr}} = 2.0$, $X_{2,\text{extr}} = 0.0$

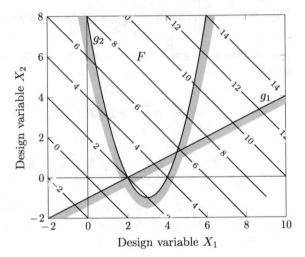

$$g_1 \leq 0 \text{ and } g_2 > 0: \qquad \Phi(X, r_p) = X_1 + X_2 + r_p\left[8 - 6X_1 + X_1^2 - X_2\right]^2, \tag{5.11}$$

$$g_1 > 0 \text{ and } g_2 \leq 0: \qquad \Phi(X, r_p) = X_1 + X_2 + r_p\left[-2 + X_1 - 2X_2\right]^2, \tag{5.12}$$

$$g_1 > 0 \text{ and } g_2 > 0: \qquad \Phi(X, r_p) = X_1 + X_2 + r_p\Big\{\left[-2 + X_1 - 2X_2\right]^2$$
$$+ \left[8 - 6X_1 + X_1^2 - X_2\right]^2\Big\}, \tag{5.13}$$

$$g_1 \leq 0 \text{ and } g_2 \leq 0: \qquad \Phi(X, r_p) = X_1 + X_2. \tag{5.14}$$

The graphical representation of the pseudo-objective function Φ for different values of the factor r_p is shown in Figs. 5.3 and 5.4.

It can be seen that the minimum (represented by the marker ●) of the pseudo-objective function Φ move closer to the minimum of the objective function F (i.e., the intersection of the dotted curves) by increasing the value of the parameter r_p.

The following Listing 5.1 shows the entire wxMaxima code for the determination of the minimum of the objective function given in Eq. (5.7).

The iteration history is illustrated in Figs. 5.5 and 5.6 for each penalty parameter, i.e. $r_{p0s} = 0.5, 1, 10, 20, 50$, and 100 and $r_{p0s} = 0.05, 0.1, 0.5, 1, 10, 20, 50$, and 100, respectively. It can be seen from this graphical representation that already the first few steps reach extremely close to the analytical minimum. However, the choice of the r_{p0s} parameters influences the initial iteration history.

To further illustrate the iteration process for a constant r_{p0s} value, let us have a closer look at the four iterations for $r_{p0s} = 0.5$, see Fig. 5.7. This figure shows the pseudo-objective function $\Phi(\alpha)$ for $r_{p0s} = 0.5$ and the initial value of the design variables $X_0 = [00]^T$. There are 16 iterations required to find the minimum based on Newton's Method (see Sect. 2.3) for this function of one variable, i.e., $\Phi = \Phi(\alpha)$.

```
(% i12)    load("my_funs.mac")$

           fpprintprec:6$
           ratprint: false$

           eps : 1/1000

           func_obj(X) := X[1] + X[2]$

           g[1](X) := -2 + X[1] - 2*X[2]$
           g[2](X) := 8 - 6*X[1] + X[1]^2 - X[2]$

           no_of_vars : 2$

           X_0 : [X[1]=0,X[2]=0]$
           alpha_0 : 1$
           r_p_0s : [0.5, 1, 10, 20, 50, 100]$

           gamma_value : 1$
           for r : 1 thru length(r_p_0s) do (
              print("=============="),
              print("For ", r_p = r_p_0s[r]),
              [X_new, pseudo_objective_fun_value] : steepest_multi_variable_constrained
                (func_obj,no_of_vars,X_0,alpha_0,eps,r_p_0s[r], gamma_value,
                "Kuhn_Tucker",false),
              print(X["1"] = rhs(X_new[1])),
              print(X["2"] = rhs(X_new[2])),
              printf(true, "The pseudo-objective function value at this point:
                    F = ~,6f", pseudo_objective_fun_value),
              X_0 : copy(X_new)
           )$
```

```
==============
For r_p=0.5
i=1 X=[X[1]=1.70618,X[2]=0.254113] func(X)=2.04843
i=2 X=[X[1]=1.77473,X[2]=-0.206132] func(X)=1.8363
i=3 X=[X[1]=1.84337,X[2]=-0.195909] func(X)=1.81754
i=4 X=[X[1]=1.84325,X[2]=-0.195087] func(X)=1.81753
Converged after 4 iterations!
X[1]=1.84325
X[2]=-0.195087
The pseudo-objective function value at this point:
F = 1.817535
==============
For r_p=1
i=1 X=[X[1]=1.84425,X[2]=-0.194087] func(X)=1.98492
i=2 X=[X[1]=1.92474,X[2]=-0.113207] func(X)=1.90695
i=3 X=[X[1]=1.91157,X[2]=-0.100063] func(X)=1.90506
i=4 X=[X[1]=1.91297,X[2]=-0.0986663] func(X)=1.90504
i=5 X=[X[1]=1.9128,X[2]=-0.0985007] func(X)=1.90504
```

Converged after 5 iterations!
X[1]=1.9128
X[2]=-0.0985007
The pseudo-objective function value at this point:
F = 1.905038
==============
For r_p=10
i=1 X=[X[1]=1.91588,X[2]=-0.0954213] func(X)=2.66732
i=2 X=[X[1]=1.99535,X[2]=-0.015239] func(X)=1.99281
i=3 X=[X[1]=1.9901,X[2]=-0.0100397] func(X)=1.99006
i=4 X=[X[1]=1.99016,X[2]=-0.00998143] func(X)=1.99006
Converged after 4 iterations!
X[1]=1.99016
X[2]=-0.00998143
The pseudo-objective function value at this point:
F = 1.990059
==============
For r_p=20
i=1 X=[X[1]=1.99116,X[2]=-0.00898127] func(X)=1.99815
i=2 X=[X[1]=1.99508,X[2]=-0.00503322] func(X)=1.99502
i=3 X=[X[1]=1.99504,X[2]=-0.00499534] func(X)=1.99501
Converged after 3 iterations!
X[1]=1.99504
X[2]=-0.00499534
The pseudo-objective function value at this point:
F = 1.995015
==============
For r_p=50
i=1 X=[X[1]=1.99654,X[2]=-0.00349521] func(X)=1.999
i=2 X=[X[1]=1.99801,X[2]=-0.00200534] func(X)=1.998
i=3 X=[X[1]=1.99801,X[2]=-0.00199922] func(X)=1.998
Converged after 3 iterations!
X[1]=1.99801
X[2]=-0.00199922
The pseudo-objective function value at this point:
F = 1.998002
==============
For r_p=100
i=1 X=[X[1]=1.99901,X[2]=-9.99205*10^-4] func(X)"=1.999
i=2 X=[X[1]=1.999,X[2]=-9.99801*10^-4] func(X)"=1.999
Converged after 2 iterations!
X[1]=1.999
X[2]=-9.99801*10^-4
The pseudo-objective function value at this point:
F = 1.999001

Module 5.1: Numerical determination of the minimum of the objective function $F(X_1, X_2)$ and the limitations due to two inequality constraints (see Eqs. (5.7)–(5.9))

Fig. 5.3 Contour diagram of the pseudo-objective function $\Phi(X_1, X_2)$: **a** $r_{p0s} = 0.05$, $X_{1,min} = 1.28917$, $X_{2,min} = -1.89695$ and **b** $r_{p0s} = 0.1$, $X_{1,min} = 1.51431$, $X_{2,min} = -0.953821$

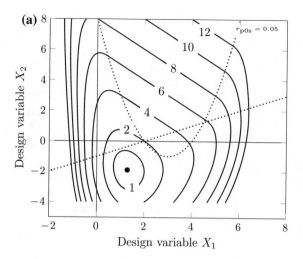

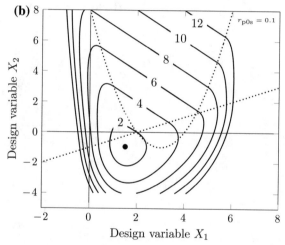

With the obtained value $\alpha_{min}^{i=1} = 0.0363018$, the first prediction for the optimized design variables can be obtained based on:

$$X^{i=1} = X_0 + \alpha_{min}^{i=1} \times S_0 \tag{5.15}$$

$$= \begin{bmatrix} 0 \\ 0 \end{bmatrix} + 0.0363018 \times \begin{bmatrix} 47 \\ 7 \end{bmatrix} \tag{5.16}$$

$$= \begin{bmatrix} 1.7061845 \\ 0.2541126 \end{bmatrix}. \tag{5.17}$$

This new value $X^{i=1}$ is used to plot again the pseudo-objective function $\Phi(\alpha)$ for the second iteration, see Fig. 5.8. The evaluation procedure is repeated and a new $\alpha_{min}^{i=2}$ is

Fig. 5.4 Contour diagram of
the pseudo-objective
function $\Phi(X_1, X_2)$: **a**
$r_{p0s} = 0.5$,
$X_{1,min} = 1.84328$, $X_{2,min} = -0.195288$ and **b** $r_{p0s} = 1.0$,
$X_{1,min} = 1.91282$, $X_{2,min} = -0.0985795$

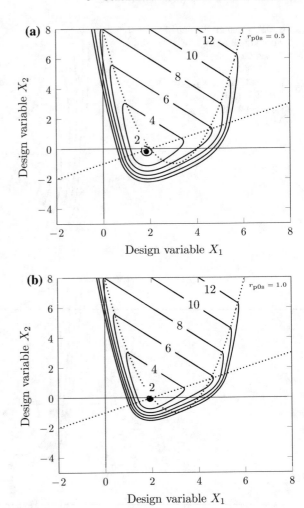

used to update the the design variable. Table 5.1 summarizes the values for the four
iterations of the case $r_{p0s} = 0.5$.

The numerical procedure continues for $r_{p0s} = 1.0$ with $X_0 = [1.8432496-0.1950870]^T$ and $\alpha_0 = 1.0$ for the first iteration step.

5.3   Supplementary Problems

5.2 Numerical Determination of the Optimal Design of a Cantilever Beam: Constant Rectangular Cross-Sectional Area

Fig. 5.5 Iteration history in the contour diagram of the objective objective function $F(X_1, X_2)$ and the limitations due to two inequality constraints (see Eqs. 5.7–5.9) for $\alpha_0 = 1.0$, $X_0 = [00]^T$, and $r_{p0s} = [0.5, 1, 10, 20, 50, 100]$. Exact solution for the minimum: $X_{1,\text{extr}} = 2.0$, $X_{2,\text{extr}} = 0.0$

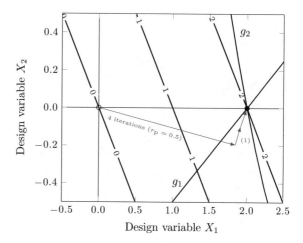

Fig. 5.6 Iteration history in the contour diagram of the objective objective function $F(X_1, X_2)$ and the limitations due to two inequality constraints (see Eqs. 5.7–5.9) for $\alpha_0 = 1.0$, $X_0 = [00]^T$, and $r_{p0s} = [0.05, 0.1, 0.5, 1, 10, 20, 50, 100]$. Exact solution for the minimum: $X_{1,\text{extr}} = 2.0$, $X_{2,\text{extr}} = 0.0$

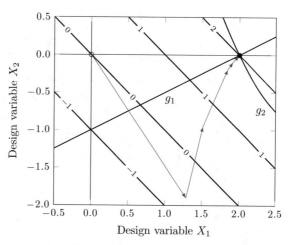

Given is a cantilever beam as shown in Fig. 5.9. The beam is loaded by a single force F_0 and has constant material (E, ϱ) and geometrical properties (I) along its axis. The material is isotropic and homogeneous and the beam theory for thin beams (Euler-Bernoulli) should be applied for this example, see [5].

Given are:

- Geometrical dimension: $L = 2540$ mm.
- Material properties of the beam: Young's modulus $E = 68948$ MPa, mass density $\varrho = 2691$ kg/m^3, initial tensile yield stress $R_{p0.2} = 247$ MPa.
- Loading: $F_0 = 2667$ N.

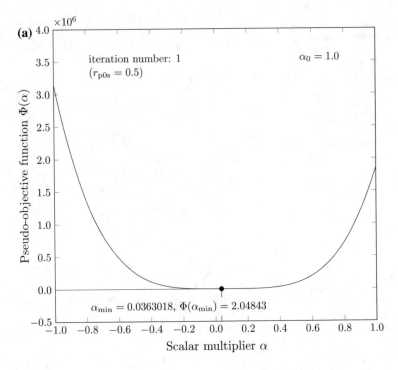

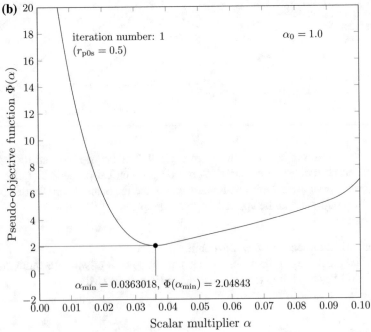

Fig. 5.7 Graphical representation of the pseudo-objective function $\Phi(\alpha)$ for $r_{p0s} = 0.5$, $X_0 = [0\,0]^T$, and the first iteration step: **a** global view and **b** magnification

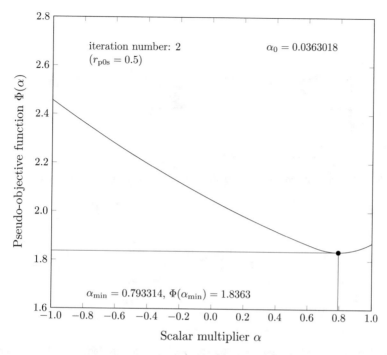

Fig. 5.8 Graphical representation of the pseudo-objective function $\Phi(\alpha)$ for $r_{p0s} = 0.5$, $X^{i=1} = [1.70618450.2541126]^T$, and the second iteration step

Table 5.1 Numerical values for the update of the design variables for the case $r_{p0s} = 0.5$, $X_0 = [00]^T$

Iteration i	X^{i-1}	S^{i-1}	α_{min}^i	X^i
1	$\begin{bmatrix} 0 \\ 0 \end{bmatrix}$	$\begin{bmatrix} 47 \\ 7 \end{bmatrix}$	0.0363018	$\begin{bmatrix} 1.7061845 \\ 0.2541126 \end{bmatrix}$
2	$\begin{bmatrix} 1.7061845 \\ 0.2541126 \end{bmatrix}$	$\begin{bmatrix} 0.0864065 \\ -0.5801540 \end{bmatrix}$	0.7933141	$\begin{bmatrix} 1.7747320 \\ -0.2061318 \end{bmatrix}$
3	$\begin{bmatrix} 1.7747320 \\ -0.2061318 \end{bmatrix}$	$\begin{bmatrix} 0.5465466 \\ 0.08140464 \end{bmatrix}$	0.1255852	$\begin{bmatrix} 1.8433701 \\ -0.1959086 \end{bmatrix}$
4	$\begin{bmatrix} 1.8433701 \\ -0.1959086 \end{bmatrix}$	$\begin{bmatrix} -5.9783264 \times 10^{-4} \\ 0.0040758 \end{bmatrix}$	0.2015744	$\begin{bmatrix} 1.8432496 \\ -0.1950870 \end{bmatrix}$

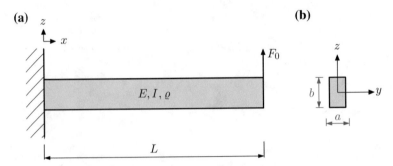

Fig. 5.9 a General configuration of the cantilever beam problem; **b** cross-sectional area

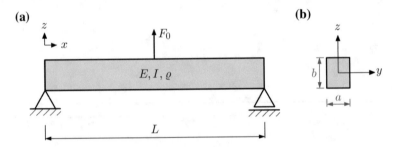

Fig. 5.10 a General configuration of the simply supported beam problem; **b** cross-sectional area

Determine the optimized cross-sectional dimensions a and b under the condition that the acting normal stress does not exceed the initial yield stress. Furthermore, the beam should not exceed a maximum deflection of $u_z(L) = r_1 L$ with $r_1 = 0.03$. Finally, the height-to-width ratio should be limited to $b \leq 20a$ to avoid instability. Use the exterior penalty function method to solve this problem.

Compare the optimized solution with the results from Problem 3.6.

5.3 Numerical Determination of the Optimal Design of a Simply Supported Beam: Constant Rectangular Cross-Sectional Area

Given is a simply supported beam as shown in Fig. 5.10. The beam is loaded in the middle by a single force F_0 and has constant material (E, ϱ) and geometrical properties (I) along its axis. The material is isotropic and homogeneous and the beam theory for thin beams (Euler-Bernoulli) should be applied for this example, see [5, 9].

Given are:

- Geometrical dimension: $L = 2540\,\text{mm}$.
- Material properties of the beam: Young's modulus $E = 68948\,\text{MPa}$, mass density $\varrho = 2691\,\text{kg/m}^3$, initial tensile yield stress $R_{p0.2} = 247\,\text{MPa}$, initial shear yield stress $\tau_p = R_{p0.2}/2$.
- Loading: $F_0 = 2667\,\text{N}$.

Fig. 5.11 a General
configuration of the short
beam problem; **b**
cross-sectional area

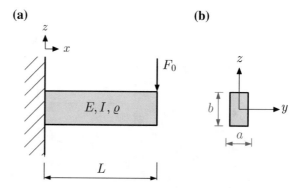

Determine the optimized cross-sectional dimensions a and b under the condition that the acting normal and shear stresses do not exceed the initial yield stresses. Furthermore, the beam should not exceed a maximum deflection of $u_z(L) = r_1 L$ with $r_1 = 0.03$ or $r_1 = 0.003$ as an alternative. Finally, the height-to-width ratio should be limited to $b \leq 20a$ to avoid instability. Use the exterior penalty function method to solve this problem.

5.4 Numerical Determination of the Optimal Design of a Short Cantilever Beam: Constant Rectangular Cross-Sectional Area

Given is a short cantilever beam as shown in Fig. 5.11. The beam is loaded by a single force F_0 and has constant material (E, ϱ) and geometrical properties (I) along its axis. The material is isotropic and homogeneous and the beam theory for thick beams should be applied for this example, see [3, 5, 6, 9] for details on the theory.

Given are:

- Geometrical dimension: $L = 846.33$ mm.
- Material properties of the beam: Young's modulus $E = 68948$ MPa, mass density $\varrho = 2691$ kg/m^3, initial tensile yield stress $R_{p0.2} = 247$ MPa.
- Loading: $F_0 = 2667$ N.

Determine the optimized cross-sectional dimensions a and b under the condition that the maximum normal *and* shear stresses do not exceed the corresponding initial yield stresses. The initial shear yield stress can be approximated based on the Tresca yield condition. Consider that the normal stress has a linear distribution while the shear stress has a parabolic distribution over the beam height. Furthermore, the height-to-width ratio should be limited to $b \leq 20a$ to avoid instability. Use the exterior penalty function method to solve this problem.

Compare the optimized solution with the results from Problem 3.8.

5.5 Optimization of a Stepped Cantilever Beam with Two Sections

The cantilever beam shown in Fig. 5.12 is loaded by a single force F_0 in negative Z-direction at its right-hand end. The beam is divided in two sections of length $L_I = L_{II} = \frac{L}{2}$. The material of each section is the same, i.e. $E_I = E_{II} = E$, but each

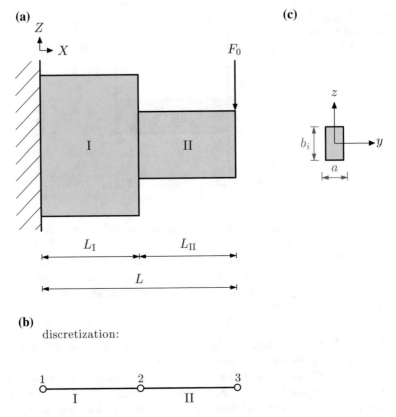

Fig. 5.12 Cantilever stepped beam with two sections: **a** general configuration, **b** discretization, and **c** cross-section of element i

cross section is different. Optimize the cross-sections, i.e. the height b_i of each section while the width of each section remains constant equal to a.

Given are:

- Geometrical dimensions: $L = 2540\,\text{mm}$, $a = 45.16\,\text{mm}$.
- Material properties of the beam: Young's modulus $E = 68948\,\text{MPa}$, mass density $\varrho = 2691\,\text{kg/m}^3$, initial tensile yield stress $R_{p0.2} = 247\,\text{MPa}$.
- Loading: $F_0 = 2667\,\text{N}$.

Determine the optimized cross-sectional dimensions b_i under the condition that the acting normal and shear stresses do not exceed the initial tensile and shear yield stresses. Furthermore, the beam should not exceed a maximum deflection of $|u_z(x = L)| = r_1 L$ with $r_1 = 0.06$. Finally, the height-to-width ratio should be limited to $b_i \leq 20a$ to avoid instability. Use the exterior penalty function method to solve this problem. Details on a finite element approach to calculate the deformation and internal reactions can be found, for example, in [4, 7, 8].

5.6 Optimization of a Stepped Simply Supported Beam with Three Sections

The simply supported beam shown in Fig. 5.13 is loaded by a single force F_0 in negative Z-direction in the middle of the beam. The beam is divided in three sections of length $L_I = L_{II} = L_{III} = \frac{L}{3}$. The material of each section is the same, i.e. $E_I = E_{II} = E_{III} = E$, but the middle section is different to the outer ones (I = III). Optimize the cross-sections, i.e. the height b_i of each section while the width of each section remains constant equal to a.

Given are:

- Geometrical dimensions: $L = 2540\,\text{mm}$, $a = 45.16\,\text{mm}$.
- Material properties of the beam: Young's modulus $E = 68948\,\text{MPa}$, mass density $\varrho = 2691\,\text{kg/m}^3$, initial tensile yield stress $R_{p0.2} = 247\,\text{MPa}$.
- Loading: $F_0 = 2667\,\text{N}$.

Determine the optimized cross-sectional dimensions b_i under the condition that the acting normal and shear stresses do not exceed the initial tesnile and shear yield stresses. Furthermore, the beam should not exceed a maximum deflection of $|u_z(x = L/2)| = r_1 L$ with $r_1 = 0.01$. Finally, the height-to-width ratio should be limited to $b_i \leq 20a$ to avoid instability. Use the exterior penalty function method to solve this problem. Details on a finite element approach to calculate the deformation and internal reactions can be found, for example, in [4, 7, 8].

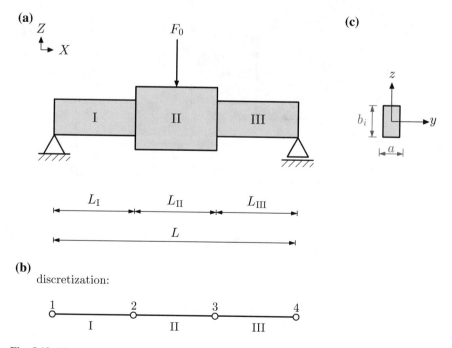

Fig. 5.13 Simply supported beam with three sections: **a** general configuration, **b** discretization, and **c** cross-section of element i

5.7 Optimization of a Stepped Cantilever Beam with Three Sections

The cantilever beam shown in Fig. 5.14 is loaded by a single force F_0 in negative Z-direction at its right-hand end. The beam is divided in three sections of length $L_{\mathrm{I}} = L_{\mathrm{II}} = L_{\mathrm{III}} = \frac{L}{3}$. The material of each section is the same, i.e. $E_{\mathrm{I}} = E_{\mathrm{II}} = E_{\mathrm{III}} = E$, but each cross section is different. Optimize the cross-sections, i.e. dimensions a_i and b_i, ...

- Geometrical dimension: $L = 2540\,\mathrm{mm}$.
- Material properties of the beam: Young's modulus $E = 68948\,\mathrm{MPa}$, mass density $\varrho = 2691\,\mathrm{kg/m^3}$, initial tensile yield stress $R_{\mathrm{p0.2}} = 247\,\mathrm{MPa}$.
- Loading: $F_0 = 2667\,\mathrm{N}$.

Determine the optimized cross-sectional dimensions a_i and b_i under the condition that the acting normal and shear stresses do not exceed the initial tensile and shear yield stresses. Furthermore, the beam should not exceed a maximum deflection of $|u_z(x = L)| = r_1 L$ with $r_1 = 0.06$. Finally, the height-to-width ratio should be lim-

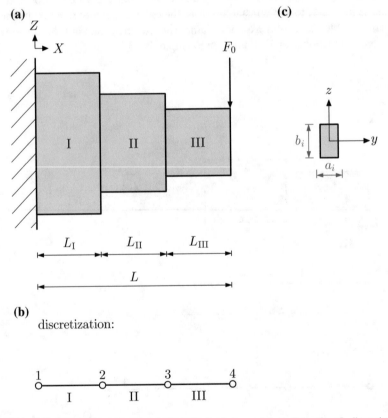

Fig. 5.14 Cantilever stepped beam with three sections: **a** general configuration, **b** discretization, and **c** cross-section of element i

ited to $b_i \leq 20a_i$ to avoid instability. Use the exterior penalty function method to solve this problem. Details on a finite element approach to calculate the deformation and internal reactions can be found, for example, in [4, 7, 8].

5.8 Optimization of a Stepped Simply Supported Beam

The simply supported beam shown in Fig. 5.15 is loaded by a single force F_0 in negative Z-direction at its middle. The beam is divided in four sections of length $L_I = \cdots = L_{IV} = \frac{L}{4}$. The material of each section is the same, i.e. $E_I = \cdots = E_{IV} = E$, but each cross section is different. Determine the deformations at each node under the assumption that the beam is symmetric, i.e $I_{IV} = I_I$, and $I_{III} = I_{II}$. Optimize the cross-sections, i.e. dimensions a_i and b_i, ...

Given are:

- Geometrical dimensions: $L = 2540$ mm.
- Material properties of the beam: Young's modulus $E = 68948$ MPa, mass density $\varrho = 2691$ kg/m^3, initial tensile yield stress $R_{p0.2} = 247$ MPa.
- Loading: $F_0 = 2667$ N.

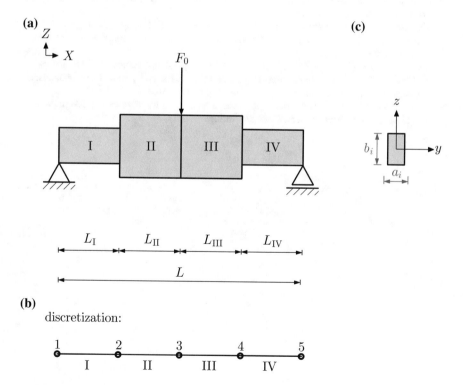

Fig. 5.15 Simply supported stepped beam with four sections: **a** general configuration, **b** discretization, and **c** cross-section of element i

Determine the optimized cross-sectional dimensions a_i and b_i under the condition that the acting normal and shear stresses do not exceed the initial tensile and shear yield stresses. Furthermore, the beam should not exceed a maximum deflection of $|u_z(x = L/2)| = r_1 L$ with $r_1 = 0.01$. Finally, the height-to-width ratio should be limited to $b_i \leq 20a_i$ to avoid instability. Use the exterior penalty function method to solve this problem. Details on a finite element approach to calculate the deformation and internal reactions can be found, for example, in [4, 7, 8].

References

1. Allen HG (1966) Optimum design of sandwich struts and beams. In: Plastics in building structures, Proceedings of a conference held in London, 14–16 June 1965. Pergamon Press, Oxford
2. Allen HG (1969) Analysis and design of structural sandwich panels. Pergamon Press, Oxford
3. Boresi AP, Schmidt RJ (2003) Advanced mechanics of materials. Wiley, New York
4. Javanbakht Z, Öchsner A (2018) Computational statics revision course. Springer, Cham
5. Öchsner A (2014) Elasto-plasticity of frame structure elements: modelling and simulation of rods and beams. Springer, Berlin
6. Öchsner A (2016) Continuum damage and fracture mechanics. Springer, Singapore
7. Öchsner A (2016) Computational statics and dynamics—an introduction based on the finite lement method. Springer, Singapore
8. Öchsner A (2018) A project-based introduction to computational statics. Springer, Cham
9. Öchsner A (2019) Leichtbaukonzepte anhand einfacher Strukturelemente: Neuer didaktischer Ansatz mit zahlreichen Übungsaufgaben. Springer Vieweg, Berlin
10. Vanderplaats GN (1999) Numerical optimization techniques for engineering design. Vanderplaats Research & Development, Colorado Springs

Chapter 6
Answers to Supplementary Problems

6.1 Problems from Chapter 2

2.8 Numerical Determination of a Minimum

The following Listing 6.1 shows the entire wxMaxima code for the determination of the minimum.

```
(% i2)    load("my_funs.mac")$
          load("engineering-format")$


(% i9)    func(x) := -2*sin(x)*(1+cos(x))$
          xmin : 0$
          xmax : %pi/2$
          eps : 0.001$

          N : round((log(eps)/(log(1-0.381966))+3))$

          print("N =", N)$
          gss(xmin, xmax, N)$

   N = 17

   K   x_min      x_1       x_2       x_max      f_min        f_1          f_2        f_max
   3  0.0000E−1 5.9999E−1 9.7081E−1 1.5708E+0  0.0000E−1  −2.0613E+0 −2.5827E+0 −2.0000E+0
   4  5.9999E−1 9.7081E−1 1.2000E+0 1.5708E+0 −2.0613E+0 −2.5827E+0 −2.5396E+0 −2.0000E+0
   5  5.9999E−1 8.2917E−1 9.7081E−1 1.2000E+0 −2.0613E+0 −2.4709E+0 −2.5827E+0 −2.5396E+0
   6  8.2917E−1 9.7081E−1 1.0583E+0 1.2000E+0 −2.4709E+0 −2.5827E+0 −2.5978E+0 −2.5396E+0
   7  9.7081E−1 1.0583E+0 1.1124E+0 1.2000E+0 −2.5827E+0 −2.5978E+0 −2.5872E+0 −2.5396E+0
```

```
 8  9.7081E-1 1.0249E+0 1.0583E+0 1.1124E+0 −2.5827E+0 −2.5968E+0 −2.5978E+0 −2.5872E+0
 9  1.0249E+0 1.0583E+0 1.0790E+0 1.1124E+0 −2.5968E+0 −2.5978E+0 −2.5955E+0 −2.5872E+0
10  1.0249E+0 1.0456E+0 1.0583E+0 1.0790E+0 −2.5968E+0 −2.5981E+0 −2.5978E+0 −2.5955E+0
11  1.0249E+0 1.0377E+0 1.0456E+0 1.0583E+0 −2.5968E+0 −2.5978E+0 −2.5981E+0 −2.5978E+0
12  1.0377E+0 1.0456E+0 1.0504E+0 1.0583E+0 −2.5978E+0 −2.5981E+0 −2.5980E+0 −2.5978E+0
13  1.0377E+0 1.0426E+0 1.0456E+0 1.0504E+0 −2.5978E+0 −2.5980E+0 −2.5981E+0 −2.5980E+0
14  1.0426E+0 1.0456E+0 1.0474E+0 1.0504E+0 −2.5980E+0 −2.5981E+0 −2.5981E+0 −2.5980E+0
15  1.0456E+0 1.0474E+0 1.0486E+0 1.0504E+0 −2.5981E+0 −2.5981E+0 −2.5981E+0 −2.5980E+0
16  1.0456E+0 1.0467E+0 1.0474E+0 1.0486E+0 −2.5981E+0 −2.5981E+0 −2.5981E+0 −2.5981E+0
17  1.0467E+0 1.0474E+0 1.0479E+0 1.0486E+0 −2.5981E+0 −2.5981E+0 −2.5981E+0 −2.5981E+0
```

Module 6.1: Numerical determination of a minimum for the function $F(X) = -2 \times \sin(X) \times (1 + \cos(X))$ in the range $0 \leq X \leq \frac{\pi}{2}$

2.9 Numerical Determination of a Maximum

In oder to find the maximum of the function $F(X)$, we solve numerically for the minimum of $-F(X)$. The following Listing 6.2 shows the entire wxMaxima code for the determination of the minimum.

```
(% i2)    load("my_funs.mac")$
          load("engineering-format")$


(% i9)    func(x) := -8*x/(x^2-2*x+4)$
          xmin : 0$
          xmax : 10$
          eps : 0.001$

          N : round((log(eps)/(log(1-0.381966))+3))$

          print("N =", N)$
          gss(xmin, xmax, N)$

N = 17
```

K	x_min	x_1	x_2	x_max	f_min	f_1	f_2	f_max
3	0.0000E-1	3.8197E+0	6.1803E+0	1.0000E+1	0.0000E-1	−2.7905E+0	−1.6572E+0	−9.5238E-1
4	0.0000E-1	2.3607E+0	3.8197E+0	6.1803E+0	0.0000E-1	−3.8927E+0	−2.7905E+0	−1.6572E+0
5	0.0000E-1	1.4590E+0	2.3607E+0	3.8197E+0	0.0000E-1	−3.6353E+0	−3.8927E+0	−2.7905E+0
6	1.4590E+0	2.3607E+0	2.9180E+0	3.8197E+0	−3.6353E+0	−3.8927E+0	−3.4953E+0	−2.7905E+0
7	1.4590E+0	2.0163E+0	2.3607E+0	2.9180E+0	−3.6353E+0	−3.9997E+0	−3.8927E+0	−3.4953E+0
8	1.4590E+0	1.8034E+0	2.0163E+0	2.3607E+0	−3.6353E+0	−3.9576E+0	−3.9997E+0	−3.8927E+0
9	1.8034E+0	2.0163E+0	2.1478E+0	2.3607E+0	−3.9576E+0	−3.9997E+0	−3.9798E+0	−3.8927E+0

10 1.8034E+0	1.9350E+0	2.0163E+0	2.1478E+0	−3.9576E+0	−3.9956E+0	−3.9997E+0	−3.9798E+0
11 1.9350E+0	2.0163E+0	2.0665E+0	2.1478E+0	−3.9956E+0	−3.9997E+0	−3.9957E+0	−3.9798E+0
12 1.9350E+0	1.9852E+0	2.0163E+0	2.0665E+0	−3.9956E+0	−3.9998E+0	−3.9997E+0	−3.9957E+0
13 1.9350E+0	1.9660E+0	1.9852E+0	2.0163E+0	−3.9956E+0	−3.9988E+0	−3.9998E+0	−3.9997E+0
14 1.9660E+0	1.9852E+0	1.9971E+0	2.0163E+0	−3.9988E+0	−3.9998E+0	−4.0000E+0	−3.9997E+0
15 1.9852E+0	1.9971E+0	2.0044E+0	2.0163E+0	−3.9998E+0	−4.0000E+0	−4.0000E+0	−3.9997E+0
16 1.9852E+0	1.9925E+0	1.9971E+0	2.0044E+0	−3.9998E+0	−3.9999E+0	−4.0000E+0	−4.0000E+0
17 1.9925E+0	1.9971E+0	1.9999E+0	2.0044E+0	−3.9999E+0	−4.0000E+0	−4.0000E+0	−4.0000E+0

Module 6.2: Numerical determination of a minimum for the function $F(X) = -\frac{8X}{X^2-2X+4}$ in the range $0 \leq X \leq 10$

2.10 Brute-Force Approach for the Determination of a Minimum

The following Listing 6.3 shows the entire wxMaxima code for the determination of the minimum based on brute-force version 1 for the particular case of $n = 10$.

```
(% i2)   load("my_funs.mac")$
         load("engineering-format")$


(% i7)   func(x) := -2*sin(x)*(1+cos(x))$
         xmin : 0$
         xmax : %pi/2$

         n : 10$

         bf_ver1(xmin, xmax, n)$

minimum lies in [ 9.4248e-1, 1.2566e+0]
X_extr = 1.0996e+0 ( i = 7 )
```

Module 6.3: Numerical determination of the minimum for the function $F(X) = -2 \times \sin(X) \times (1 + \cos(X))$ in the range $0 \leq X \leq \frac{\pi}{2}$ based on the brute-force approach (version 1) for $n = 10$ (exact value: $X_{\text{extr}} = 1.047$)

Table 6.1 Summary of detected minimum values (exact value: $X_{extr} = 1.047$) for different parameters n, i.e. different step sizes (brute-force version 1)

n	X_{min}	X_{max}	X_{extr}	i
4	7.8540e−1	1.5708e+0	1.1781e+0	3
8	7.8540e−1	1.1781e+0	9.8175e−1	5
10	9.4248e−1	1.2566e+0	1.0996e+0	7
15	9.4248e−1	1.1519e+0	1.0472e−0	10

Table 6.2 Summary of detected minimum values (exact value: $X_{extr} = 1.047$) for different parameters n, i.e. different step sizes, and different start values X_0 (brute-force version 2)

n	X_{min}	X_{max}	X_{extr}	i
$X_0 = 0.2$				
10	9.8540e−1	1.1425e+0	1.0639e+0	6
15	1.0378e+0	1.1425e+0	1.0901e+0	9
20	1.0639e+0	1.1425e+0	1.1032e+0	12
25	1.0168e+0	1.0796e+0	1.0482e+0	14
30	1.0378e+0	1.0901e+0	1.0639e+0	17
35	1.0527e+0	1.0976e+0	1.0752e+0	20
40	1.0639e+0	1.1032e+0	1.0836e+0	23
1000	1.0467e+0	1.0482e+0	1.0474e+0	540
$X_0 = 1.3$				
10	9.8584e−1	1.1429e+0	1.0644e+0	3
15	1.0906e+0	1.1953e+0	1.1429e+0	3
20	1.0644e+0	1.1429e+0	1.1037e+0	4
25	1.0487e+0	1.1115e+0	1.0801e+0	5
30	1.0382e+0	1.0906e+0	1.0644e+0	6
35	1.0307e+0	1.0756e+0	1.0532e+0	7
40	1.0644e+0	1.1037e+0	1.0840e+0	7
1000	1.0471e+0	1.0487e+0	1.0479e+0	162

Other values for a variation of the parameter n are summarized in Table 6.1.

The wxMaxima code for the application of brute-force method 2 is illustrated in Listing 6.4.

Other values for a variation of the parameter n and the start value X_0 are summarized in Table 6.2.

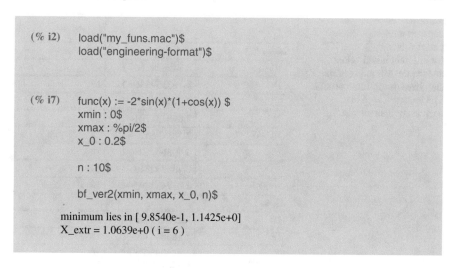

```
(% i2)    load("my_funs.mac")$
          load("engineering-format")$


(% i7)    func(x) := -2*sin(x)*(1+cos(x)) $
          xmin : 0$
          xmax : %pi/2$
          x_0 : 0.2$

          n : 10$

          bf_ver2(xmin, xmax, x_0, n)$

          minimum lies in [ 9.8540e-1, 1.1425e+0]
          X_extr = 1.0639e+0 ( i = 6 )
```

Module 6.4: Numerical determination of the minimum for the function $F(X) = -2 \times \sin(X) \times (1 + \cos(X))$ in the range $0 \le X \le \frac{\pi}{2}$ based on the brute-force approach (version 2) for $n = 10$ and $X_0 = 0.2$ (exact value: $X_{\text{extr}} = 1.047$).

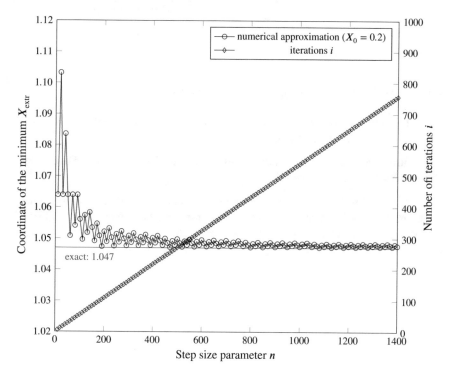

Fig. 6.1 Convergence rate of the brute-force approach (version 2) to detect the minimum of the function $F(X) = -2 \times \sin(X) \times (1 + \cos(X))$ in the range $0 \le X \le \frac{\pi}{2}$

Table 6.3 Summary of detected minimum values (exact value: $X_{extr} = 1.047$) for different initial values X_0 and parameters n, i.e. initial step sizes (brute-force version 3). Case: $\alpha^{(i)} = \frac{1}{10}$, i.e. decreasing interval size

n	X_{extr}	i
$X_0 = 0.2$		
10	3.74532925e$-$1	17
15	3.16355283e$-$1	17
20	2.87266463e$-$1	17
25	2.69813170e$-$1	17
30	2.58177642e$-$1	17
35	2.49866550e$-$1	17
40	2.43633231e$-$1	17
1000	2.01745329e$-$1	16
$X_0 = 1.3$		
10	1.12546707e$+$0	16
15	1.18364472e$+$0	16
20	1.21273354e$+$0	16
25	1.23018683e$+$0	16
30	1.24182236e$+$0	15
35	1.25013345e$+$0	16
40	1.25636677e$+$0	16
1000	1.29825467e$+$0	14

2.11 Convergence Rate for the Brute-Force Approac

Fig. 6.1 shows the convergence rate expressed as minimum coordinate X_{extr} as a function of step size parameter n.

2.12 Numerical Determination of a Minimum Based on the Brute-Force Algorithm with Variable Interval Size

The following Listing 6.5 shows the entire wxMaxima code for the determination of the minimum based on brute-force version 3 for the function given in Eq. (2.23) and $\alpha^{(i)} = 1/10$. In case that the Fibonacci sequence should be used, the code must me modified by alpha: "Fibonacci" (Tables 6.3, 6.4 and 6.5).

```
(% i2)    load("my_funs.mac")$
          load("engineering-format")$


(% i9)    func(x) := -2*sin(x)*(1+cos(x)),
          xmin : 0,
          xmax : %pi/2,
          x0 : [0.2, 1.3],
          alpha : 1/10,
          n : [10, 15, 20, 25, 30, 35, 40, 1000],
          for i : 1 thru length(x0) do (
              printf(true, "~% X_0 = ~f", x0[i]),
              for j : 1 thru length(n) do (
              bf_ver2_varN_table(xmin, xmax, x0[i], n[j], alpha)
              )
          )$


       X_0 = 0.2
          10 3.74532925e-1 17
          15 3.16355283e-1 17
          20 2.87266463e-1 17
          25 2.69813170e-1 17
          30 2.58177642e-1 17
          35 2.49866550e-1 17
          40 2.43633231e-1 17
        1000 2.01745329e-1 16
       X_0 = 1.3
          10 1.12546707e+0 16
          15 1.18364472e+0 16
          20 1.21273354e+0 16
          25 1.23018683e+0 16
          30 1.24182236e+0 15
          35 1.25013345e+0 16
          40 1.25636677e+0 16
        1000 1.29825467e+0 14
```

Module 6.5: Numerical determination of the minimum for the function $F(X) = -2 \times \sin(X) \times (1 + \cos(X))$ in the range $0 \leq X \leq \frac{\pi}{2}$ based on the brute-force approach with variable step size (version 3) and $\alpha^{(i)} = 1/10$ (exact value: $X_{\text{extr}} = 1.047$).

Table 6.4 Summary of detected minimum values (exact value: $X_{\text{extr}} = 1.047$) for different initial values X_0 and parameters n, i.e. initial step sizes (brute-force version 3). Case: $\alpha^{(i)} = 1.5$, i.e. increasing interval size

n	X_{extr}	i
$X_0 = 0.2$		
10	1.21120014e+0	4
15	1.04952265e+0	6
20	1.09481826e+0	13
25	1.26715975e+0	6
30	1.08929980e+0	6
35	1.08465450e+0	12
40	1.23973218e+0	7
1000	1.07091944e+0	20
$X_0 = 1.3$		
10	1.26073009e+0	2
15	1.15601034e+0	3
20	1.19200775e+0	3
25	1.10757745e+0	4
30	1.13964787e+0	4
35	1.16255532e+0	4
40	1.17973591e+0	4
1000	1.09933652e+0	12

2.13 Application of the Principle of Minimum Energy to a Linear Spring Problem

The total potential energy (Π), i.e. the sum of the strain energy (Π_i) and the work done by the external loads (Π_e), can be written as (see Fig. 6.2):

$$\Pi = \Pi_i + \Pi_e \tag{6.1}$$

$$= \frac{1}{2} k X^2 - F_0 X . \tag{6.2}$$

A minimum of the total potential energy requires that

$$\frac{\partial \Pi}{\partial X} = k X - F_0 \overset{!}{=} 0 , \tag{6.3}$$

which results in the following *analytical* solution for the minimum:

$$X_{\text{extr}} = \frac{F_0}{k} = 0.625 \, \text{mm} . \tag{6.4}$$

Table 6.5 Summary of detected minimum values (exact value: $X_{\text{extr}} = 1.047$) for different initial values X_0 and parameters n, i.e. initial step sizes (brute-force version 3). Case: $\alpha^{(i)} = 1, 1, 2, 3, 5, 8, 13, \ldots$, i.e. increasing interval size based on Fibonacci sequence

n	X_{extr}	i
$X_0 = 0.2$		
10	1.23044239e+0	7
15	1.10932478e+0	11
20	1.13462381e+0	7
25	1.24703000e+0	9
30	1.10519756e+0	9
35	1.05042608e+0	11
40	1.05128679e+0	10
1000	1.22872317e+0	14
$X_0 = 1.3$		
10	1.06438055e+0	3
15	1.14292037e+0	3
20	1.14292037e+0	4
25	1.17433629e+0	5
30	1.19528024e+0	5
35	1.21024021e+0	5
40	1.22146018e+0	5
1000	1.34748831e+0	10

Fig. 6.2 Graphical representation of the objective function, i.e., the total potential energy ($F = \Pi$). Exact solution for the minimum: $X_{\text{extr}} = 0.625$ mm

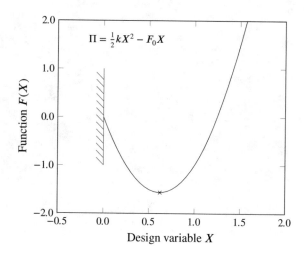

The following Listing 6.6 shows the entire wxMaxima code for the *numerical* determination of the minimum.

```
(% i2)    load("my_funs.mac")$
          load("engineering-format")$


(% i11)   k : 8$
          F_0 : 5$

          func(X) := (1/2)*k*X^2 - F_0*X$
          Xmin : 0$
          Xmax : 2$
          eps : 0.0001$

          N : round((log(eps)/(log(1-0.381966))+3))$

          print("N =", N)$
          gss(xmin, xmax, N)$
```

N = 22

K	x_min	x_1	x_2	x_max	f_min	f_1	f_2	f_max
3	0.0000E+0	7.6393E−1	1.2361E+0	2.0000E+0	0.0000E+0	−1.4853E+0	−6.8884E−2	6.0000E+0
4	0.0000E+0	4.7214E−1	7.6393E−1	1.2361E+0	0.0000E+0	−1.4690E+0	−1.4853E+0	−6.8884E−2
5	4.7214E−1	7.6393E−1	9.4427E−1	1.2361E+0	−1.4690E+0	−1.4853E+0	−1.1548E+0	−6.8884E−2
6	4.7214E−1	6.5248E−1	7.6393E−1	9.4427E−1	−1.4690E+0	−1.5595E+0	−1.4853E+0	−1.1548E+0
7	4.7214E−1	5.8359E−1	6.5248E−1	7.6393E−1	−1.4690E+0	−1.5556E+0	−1.5595E+0	−1.4853E+0
8	5.8359E−1	6.5248E−1	6.9505E−1	7.6393E−1	−1.5556E+0	−1.5595E+0	−1.5429E+0	−1.4853E+0
9	5.8359E−1	6.2616E−1	6.5248E−1	6.9505E−1	−1.5556E+0	−1.5625E+0	−1.5595E+0	−1.5429E+0
10	5.8359E−1	6.0990E−1	6.2616E−1	6.5248E−1	−1.5556E+0	−1.5616E+0	−1.5625E+0	−1.5595E+0
11	6.0990E−1	6.2616E−1	6.3621E−1	6.5248E−1	−1.5616E+0	−1.5625E+0	−1.5620E+0	−1.5595E+0
12	6.0990E−1	6.1995E−1	6.2616E−1	6.3621E−1	−1.5616E+0	−1.5624E+0	−1.5625E+0	−1.5620E+0
13	6.1995E−1	6.2616E−1	6.3000E−1	6.3621E−1	−1.5624E+0	−1.5625E+0	−1.5624E+0	−1.5620E+0
14	6.1995E−1	6.2379E−1	6.2616E−1	6.3000E−1	−1.5624E+0	−1.5625E+0	−1.5625E+0	−1.5624E+0
15	6.2379E−1	6.2616E−1	6.2763E−1	6.3000E−1	−1.5625E+0	−1.5625E+0	−1.5625E+0	−1.5624E+0
16	6.2379E−1	6.2526E−1	6.2616E−1	6.2763E−1	−1.5625E+0	−1.5625E+0	−1.5625E+0	−1.5625E+0
17	6.2379E−1	6.2470E−1	6.2526E−1	6.2616E−1	−1.5625E+0	−1.5625E+0	−1.5625E+0	−1.5625E+0
18	6.2470E−1	6.2526E−1	6.2560E−1	6.2616E−1	−1.5625E+0	−1.5625E+0	−1.5625E+0	−1.5625E+0
19	6.2470E−1	6.2504E−1	6.2526E−1	6.2560E−1	−1.5625E+0	−1.5625E+0	−1.5625E+0	−1.5625E+0
20	6.2470E−1	6.2491E−1	6.2504E−1	6.2526E−1	−1.5625E+0	−1.5625E+0	−1.5625E+0	−1.5625E+0
21	6.2491E−1	6.2504E−1	6.2513E−1	6.2526E−1	−1.5625E+0	−1.5625E+0	−1.5625E+0	−1.5625E+0
22	6.2491E−1	6.2499E−1	6.2504E−1	6.2513E−1	−1.5625E+0	−1.5625E+0	−1.5625E+0	−1.5625E+0

Module 6.6: Numerical determination of the minimum for the function $F(X) = \frac{1}{2}kX^2 - F_0 X$ in the range $0 \leq X \leq 2$ (exact value: $X_{extr} = 0.625$)

6.2 Problems from Chapter 3

3.6 Numerical Determination of the Optimal Design of a Cantilever Beam

The objective function, i.e. the mass of the beam, can be stated as a function of the design variable $X = a$ as:

$$F(X) = m(X) = 2\varrho L X^2, \tag{6.5}$$

which is to be minimized under the following two inequality constraints (see Fig. 6.3), [1]:

$$g_1(X) = \frac{F_0 L^3}{2EX^4} - r_1 L \leq 0 \quad \text{(displacement)}, \tag{6.6}$$

$$g_2(X) = \frac{3F_0 L}{2X^3} - R_{p0.2} \leq 0 \quad \text{(stress)}. \tag{6.7}$$

The following Listing 6.7 shows the entire wxMaxima code for the determination of the minimum for the set of functions given in Eqs. (6.5)–(6.7).

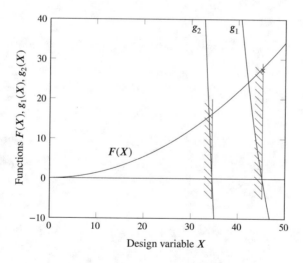

Fig. 6.3 Optimal design of a cantilever beam: $L = 2540$ mm, $F_0 = 2667$ N, $E = 68948$ MPa, $R_{p0.2} = 247$ MPa, $\varrho = 2.691 \times 10^{-6}$ kg/mm³, $r_1 = 0.03$ (original set of equations (6.5)–(6.7)). Exact solution for the minimum: $X_{\text{extr}} = 45.160$ mm

(% i3)   load("my_funs.mac")$
           load(to_poly_solve)$ /* to check if all the roots are real (isreal_p(X))) */
           ratprint : false$

(% i28)   len : 2540$
           F_0 : 2667$
           Em : 68948$
           R_p02 : 247$
           ro : 2.691E-6$
           r_1 : 0.03$

           f(X) := 2*ro*len*(X^2)$
           g[1](X) := ((F_0*(len^3)) / (2*Em*(X^4))) - (r_1*len)$
           g[2](X) := ((3*F_0*len) / (2*(X^3))) - R_p02$

           Xmin : 0$
           Xmax : 50$
           X0 : 20$
           alpha : 1$
           n : 1000$

           r_p_list : [1, 10, 50, 100, 120, 600]$
           gamma : 1$

           print("==============================")$
           print("========== Solution ==========")$
           print("==============================")$
           print(" ")$
           print("The pseudo-objective function for different ranges of X:")$
           constrained_one_variable_range_detection()$
           print(" ")$
           print("==============================")$
           for i:1 thru length(r_p_list) do (
               r_p_0 : r_p_list[i],
               print(" "),
               printf(true, "~% For r_p = ~,6f :", r_p_0),
               print(" "),
               X_extr : one_variable_constrained_exterior_penalty(Xmin, Xmax, X0, n,
                   alpha, r_p_0, gamma),
               print(" "),
               printf(true, "~% value of the non-penalized function at X = ~,6f : ~,6f", X_extr,
                                 f(X_extr)),
               printf(true, "~% value of the penalized function at X = ~,6f : ~,6f", X_extr,
                                 func(X_extr)),
               print(" "),
               print("==============================")
)$

```
===============================
========= Solution =========
===============================
```

The pseudo-objective function for different ranges of X:

For 0.000000 < X < 34.521025 :

$$\Phi = 0.01367028X^2 + r_p\left(\left(\frac{10161270}{X^3} - 247\right)^2 + \left(\frac{5463037461000}{17237X^4} - 76.2\right)^2\right)$$

For 34.521025 < X < 45.160047 :

$$\Phi = 0.01367028X^2 + r_p\left(\frac{5463037461000}{17237X^4} - 76.2\right)^2$$

For 45.160047 < X < 50.000000 :

$$\Phi = 0.01367028X^2$$

```
===============================
```

For r_p = 1.000000 :

r_p: 1.000000 , X_extr: 45.175000 , Number of iterations: 504

value of the non-penalized function at X = 45.175000 : 27.898043
value of the penalized function at X = 45.175000 : 27.898043

```
===============================
```

For r_p = 10.000000 :

r_p: 10.000000 , X_extr: 45.175000 , Number of iterations: 504

value of the non-penalized function at X = 45.175000 : 27.898043
value of the penalized function at X = 45.175000 : 27.898043

```
===============================
```

For r_p = 50.000000 :

r_p: 50.000000 , X_extr: 45.225000 , Number of iterations: 505

value of the non-penalized function at X = 45.225000 : 27.959832
value of the penalized function at X = 45.225000 : 27.959832

===============================

For r_p = 100.000000 :

r_p: 100.000000 , X_extr: 45.225000 , Number of iterations: 505
value of the non-penalized function at X = 45.225000 : 27.959832
value of the penalized function at X = 45.225000 : 27.959832

===============================

For r_p = 120.000000 :

r_p: 120.000000 , X_extr: 45.225000 , Number of iterations: 505

value of the non-penalized function at X = 45.225000 : 27.959832
value of the penalized function at X = 45.225000 : 27.959832

===============================

For r_p = 600.000000 :

r_p: 600.000000 , X_extr: 45.225000 , Number of iterations: 505

value of the non-penalized function at X = 45.225000 : 27.959832
value of the penalized function at X = 45.225000 : 27.959832

===============================

Module 6.7: Numerical determination of the minimum for the function (6.5) under consideration of the inequality constraints g_1 and g_2 in the range $0 \leq X \leq 50$ based on the exterior penalty function method (exact solution: $X_{\text{extr}} = 45.16$)

Analytical solution based on graphical representation in Fig. 6.3:

$$X_{\text{extr}} = \left(\frac{F_0 L^3}{2 E r_1 L} \right)^{\frac{1}{4}} = 45.16 \, \text{mm} . \tag{6.8}$$

Alternative to Eqs. (6.5)–(6.7), one may prefer a normalized representation, i.e. without a particular unit. Introducing the normalized geometric dimension $a^n = a/L = X^n$, one can state the corresponding equations for the optimization as:

Fig. 6.4 Optimal design of a cantilever beam: **a** normalized equations and **b** normalized equations with scaled inequality constraints ($\beta = 7000$). Exact solution for the minimum: $X^n_{extr} = 0.01778$

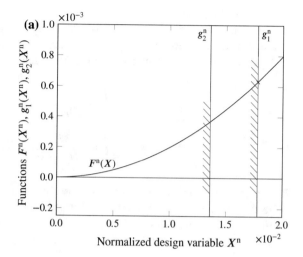

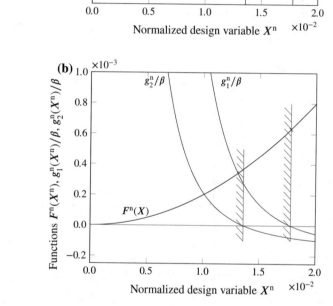

$$F^n(X^n) = F^n(a^n) = \frac{m}{\varrho \times L^3} = 2 \times (a^n)^2 = 2 \times (X^n)^2 \,, \tag{6.9}$$

$$g_1^n(X^n) = g_1^n(a^n) = \frac{F_0}{2EL^2 r_1(a^n)^4} - 1 = \frac{F_0}{2EL^2 r_1(X^n)^4} - 1 \le 0 \,, \tag{6.10}$$

$$g_2^n(X^n) = g_2^n(a^n) = \frac{3F_0}{2L^2 R_{p0.2}(a^n)^3} - 1 = \frac{3F_0}{2L^2 R_{p0.2}(X^n)^3} - 1 \le 0 \,. \tag{6.11}$$

The graphical representation of these normalized equations is given in Fig. 6.4a. One can see that both inequality constraints have again a single root in the chosen range of the abscissa. However, the slopes of both constraints are extremely steep. Thus, a scaling factor β can be introduced to modify the inequality constraints as follows:

$$g_1^n(X^n)/\beta = \frac{1}{\beta}\left(\frac{F_0}{2EL^2r_1(X^n)^4} - 1 \leq 0\right),$$
(6.12)

$$g_2^n(X^n)/\beta = \frac{1}{\beta}\left(\frac{3F_0}{2L^2R_{p0.2}(X^n)^3} - 1 \leq 0\right).$$
(6.13)

It is obvious that such a scaling does not influence the locations of the roots while a more similar representation of all curves is provided, see Fig. 6.4b.

Listing 6.8 shows the entire wxMaxima code for the determination of the minimum for the set of the *normalized* functions given in Eqs. (6.9), (6.12) and (6.13).

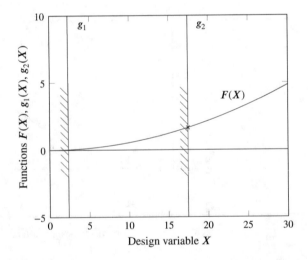

Fig. 6.5 Optimal design of a compression strut: $L = 1000\,\text{mm}$, $F_0 = 2667\,\text{N}$, $E = 70000\,\text{MPa}$, $R_{p0.2} = 247\,\text{MPa}$, $\varrho = 2.691 \times 10^{-6}\,\text{kg/mm}^3$ (considered set of equations (6.14)–(6.16)). Exact solution for the minimum: $X_{\text{extr}} = 17.447\,\text{mm}$

```
(% i3)   load("my_funs.mac")$
         load(to_poly_solve)$ /* to check if all the roots are real (isreal_p(X))) */
         ratprint : false$



(% i29)  len : 2540$
         F_0 : 2667$
         Em : 68948$
         R_p02 : 247$
         ro : 2.691E-6$
         r_1 : 0.03$
         beta : 7000$

         f(X) := 2*(X^2)$
         g[1](X) := (1/beta)*(( F_0 / (2*Em*(len^2)*r_1*(X^4)) ) - 1)$
         g[2](X) := (1/beta)*(( (3*F_0) / (2*(len^2)*R_p02*(X^3)) ) - 1)$

         Xmin : 0$
         Xmax : 2E-2$
         X0 : 5E-3$
         alpha : 1$
         n : 1000$

         r_p_list : [1, 10, 50, 100, 120, 600]$
         gamma : 1$

         print("==============================")$
         print("========== Solution ==========")$
         print("==============================")$
         print(" ")$
         print("The pseudo-objective function for different ranges of X:")$
         constrained_one_variable_range_detection()$
         print(" ")$
         print("==============================")$
         for i:1 thru length(r_p_list) do (
             r_p_0 : r_p_list[i],
             print(" "),
             printf(true, "~% For r_p = ~,6f :", r_p_0),
             print(" "),
             X_extr : one_variable_constrained_exterior_penalty(Xmin, Xmax, X0, n,
                 alpha, r_p_0, gamma),
             print(" "),
             r_p : copy(r_p_0),
             printf(true, "~% value of the non-penalized function at X = ~,6f : ~,6f", X_extr,
                             f(X_extr)),
             printf(true, "~% value of the penalized function at X = ~,6f : ~,6f", X_extr,
                             func(X_extr)),
             print(" "),
             print("==============================")
             X0 : copy(X_extr)
         )$
```

```
==============================
========= Solution ==========
==============================
```

The pseudo-objective function for different ranges of X:

For 0.000000 < X < 0.010787 :

$$\Phi = 2X^2 + r_p \left(\frac{\left(\frac{63}{50190400X^3} - 1 \right)^2}{49000000} + \frac{\left(\frac{9.99269562500371110^{-8}}{X^4} - 1 \right)^2}{49000000} \right)$$

For 0.010787 < X < 0.017780 :

$$\Phi = 2X^2 + \frac{r_p \left(\frac{9.99269562500371110^{-8}}{X^4} - 1 \right)^2}{49000000}$$

For 0.017780 < X < 0.020000 :

$$\Phi = 2X^2$$

```
==============================
```

For r_p = 1.000000 :

r_p: 1.000000 , X_extr: 0.007250 , Number of iterations: 113

value of the non-penalized function at X = 0.007250 : 0.000105
value of the penalized function at X = 0.007250 : 0.000130

```
==============================
```

For r_p = 10.000000 :

r_p: 10.000000 , X_extr: 0.009100 , Number of iterations: 93

value of the non-penalized function at X = 0.009100 : 0.000166
value of the penalized function at X = 0.009100 : 0.000203

```
===============================

For r_p = 50.000000 :

r_p: 50.000000 , X_extr: 0.010610 , Number of iterations: 76
value of the non-penalized function at X = 0.010610 : 0.000225
value of the penalized function at X = 0.010610 : 0.000274


===============================

For r_p = 100.000000 :

r_p: 100.000000 , X_extr: 0.011320 , Number of iterations: 36

value of the non-penalized function at X = 0.011320 : 0.000256
value of the penalized function at X = 0.011320 : 0.000309


===============================

For r_p = 120.000000 :

r_p: 120.000000 , X_extr: 0.011510 , Number of iterations: 10

value of the non-penalized function at X = 0.011510 : 0.000265
value of the penalized function at X = 0.011510 : 0.000319


===============================
For r_p = 600.000000 :

r_p: 600.000000 , X_extr: 0.013280 , Number of iterations: 89

value of the non-penalized function at X = 0.013280 : 0.000353
value of the penalized function at X = 0.013280 : 0.000413


===============================
```

Module 6.8: Numerical determination of the minimum for the function (6.9) under consideration of the normalized inequality constraints g_1^n and g_2^n in the range $0 \leq X^n \leq 0.02$ based on the exterior penalty function method (exact solution: $X_{extr}^n = 0.01778$)

3.7 Numerical Determination of the Optimal Design of a Compression Strut

The objective function, i.e. the mass of the strut, can be stated as a function of the design variable $X = a$ as:

$$F(X) = m(X) = 2\varrho L X^2 , \qquad (6.14)$$

which is to be minimized under the following two inequality constraints (see Fig. 6.5), [1]:

$$g_1(X) = \frac{F_0}{2X^2} - R_{p0.2} \leq 0 \quad \text{(stress)},\tag{6.15}$$

$$g_2(X) = F_0 - \frac{\pi^2 E X^4}{24L^2} \leq 0 \quad \text{(buckling)}.\tag{6.16}$$

The following Listing 6.9 shows the entire wxMaxima code for the determination of the minimum for the set of functions given in Eqs. (6.14)–(6.16).

```
(% i3)   load("my_funs.mac")$
         load(to_poly_solve)$ /* to check if all the roots are real (isreal_p(X))) */
         ratprint : false$


(% i28)  len : 1000$
         F_0 : 2667$
         Em : 70000$
         R_p02 : 247$
         ro : 2.691E-6$
         r_1 : 0.03$

         f(X) := 2*ro*len*(X^2)$
         g[1](X) := (F_0 / (2*(X^2)) ) - R_p02$
         g[2](X) := F_0 - ( (%pi^2*Em*X^4) / (24*(len^2)) )$

         Xmin : 0$
         Xmax : 30$
         X0 : 10$
         alpha : 1$
         n : 1000$

         r_p_list : [1, 10, 50, 100, 120, 600]$
         gamma : 1$

         print("==============================")$
         print("========== Solution ==========")$
         print("==============================")$
         print(" ")$
         print("The pseudo-objective function for different ranges of X:")$
         constrained_one_variable_range_detection()$
         print(" ")$
         print("==============================")$
         for i:1 thru length(r_p_list) do (
```

```
    r_p_0 : r_p_list[i],
    print(" "),
    printf(true, "~% For r_p = ~,6f :", r_p_0),
    print(" "),
    X_extr : one_variable_constrained_exterior_penalty(Xmin, Xmax, X0, n,
        alpha, r_p_0, gamma),
    print(" "),
    r_p : copy(r_p_0),
    printf(true, "~% value of the non-penalized function at X = ~,6f : ~,6f", X_extr,
                    f(X_extr)),
    printf(true, "~% value of the penalized function at X = ~,6f : ~,6f", X_extr,
                    func(X_extr)),
    print(" "),
    print("==============================")
    X0 : copy(X_extr)
)$
```

```
==============================
========= Solution ==========
==============================
```

The pseudo-objective function for different ranges of X:

For 0.000000 < X < 2.323529 :

$$\Phi = \left(\left(2667 - \frac{7\pi^2 X^4}{2400} \right)^2 + \left(\frac{2667}{2X^2} - 247 \right)^2 \right) r_p + 0.005382 X^2$$

For 2.323529 < X < 17.446532 :

$$\Phi = \left(2667 - \frac{7\pi^2 X^4}{2400} \right)^2 r_p + 0.005382 X^2$$

For 17.446532 < X < 30.000000 :

$$\Phi = 0.005382 X^2$$
```
==============================
```

For r_p = 1.000000 :

r_p: 1.000000 , X_extr: 17.485000 , Number of iterations: 250

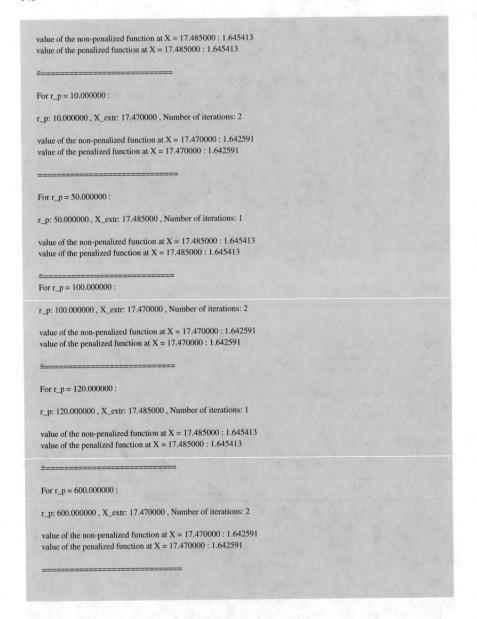

value of the non-penalized function at X = 17.485000 : 1.645413
value of the penalized function at X = 17.485000 : 1.645413

==============================

For r_p = 10.000000 :

r_p: 10.000000 , X_extr: 17.470000 , Number of iterations: 2

value of the non-penalized function at X = 17.470000 : 1.642591
value of the penalized function at X = 17.470000 : 1.642591

==============================

For r_p = 50.000000 :

r_p: 50.000000 , X_extr: 17.485000 , Number of iterations: 1

value of the non-penalized function at X = 17.485000 : 1.645413
value of the penalized function at X = 17.485000 : 1.645413

==============================

For r_p = 100.000000 :

r_p: 100.000000 , X_extr: 17.470000 , Number of iterations: 2

value of the non-penalized function at X = 17.470000 : 1.642591
value of the penalized function at X = 17.470000 : 1.642591

==============================

For r_p = 120.000000 :

r_p: 120.000000 , X_extr: 17.485000 , Number of iterations: 1

value of the non-penalized function at X = 17.485000 : 1.645413
value of the penalized function at X = 17.485000 : 1.645413

==============================

For r_p = 600.000000 :

r_p: 600.000000 , X_extr: 17.470000 , Number of iterations: 2

value of the non-penalized function at X = 17.470000 : 1.642591
value of the penalized function at X = 17.470000 : 1.642591

==============================

Module 6.9: Numerical determination of the minimum for the function (6.14) under consideration of the inequality constraints g_1 and g_2 in the range $0 \leq X \leq 30$ based on the exterior penalty function method (exact solution: $X_{extr} = 17.447$)

3.8 Numerical Determination of the Optimal Design of a Short Cantilever Beam

The objective function, i.e. the mass of the beam, can be stated as a function of the design variable $X = a$ as:

$$F(X) = m(X) = 2\varrho L X^2,\tag{6.17}$$

which is to be minimized under the following two inequality constraints (see Fig. 6.6), [1]:

$$g_1(X) = \frac{3F_0 L}{2X^3} - R_{p0.2} \le 0 \quad \text{(normal stress)},\tag{6.18}$$

$$g_2(X) = \frac{3F_0}{4X^2} - \frac{R_{p0.2}}{2} \le 0 \quad \text{(shear stress)}.\tag{6.19}$$

The following Listing 6.10 shows the entire wxMaxima code for the determination of the minimum for the set of functions given in Eqs. (6.17)–(6.19).

Fig. 6.6 Optimal design of a short cantilever beam: $L = 846.67$ mm, $F_0 = 2667$ N, $E = 68948$ MPa, $R_{p0.2} = 247$ MPa, $\varrho = 2.691 \times 10^{-6}$ kg/mm^3, (set of equations (6.17)–(6.19)). Exact solution for the minimum: $X_{\text{extr}} = 23.936$ mm

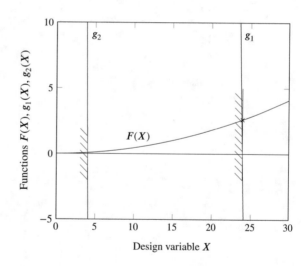

```
(% i3)    load("my_funs.mac")$
          load(to_poly_solve)$ /* to check if all the roots are real (isreal_p(X))) */
          ratprint : false$


(% i28)   len : 846.67$
          F_0 : 2667$
          Em : 68948$
          R_p02 : 247$
          ro : 2.691E-6$
          r_1 : 0.03$

          f(X) := 2*ro*len*(X^2)$
          g[1](X) := (3*F_0*len / (2*(X^3)) ) - R_p02$
          g[2](X) := 3*F_0/(4*X^2) - ( R_p02 / 2 )$

          Xmin : 0$
          Xmax : 30$
          X0 : 10$
          alpha : 1$
          n : 1000$

          r_p_list : [1, 10, 50, 100, 120, 600]$
          gamma : 1$

          print("==============================")$
          print("========== Solution ==========")$
          print("==============================")$
          print(" ")$
          print("The pseudo-objective function for different ranges of X:")$
          constrained_one_variable_range_detection()$
          print(" ")$
          print("==============================")$
          for i:1 thru length(r_p_list) do (
              r_p_0 : r_p_list[i],
              print(" "),
              printf(true, "~% For r_p = ~,6f :", r_p_0),
              print(" "),
              X_extr : one_variable_constrained_exterior_penalty(Xmin, Xmax, X0, n,
                  alpha, r_p_0, gamma),
              print(" "),
              r_p : copy(r_p_0),
              printf(true, "~% value of the non-penalized function at X = ~,6f : ~,6f", X_extr,
                              f(X_extr)),
              printf(true, "~% value of the penalized function at X = ~,6f : ~,6f", X_extr,
                              func(X_extr)),
              print(" "),
              print("==============================")
              X0 : copy(X_extr)
          )$
```

```
===============================
========== Solution ==========
===============================
```

The pseudo-objective function for different ranges of X:

For 0.000000 < X < 4.024470 :

$$\Phi = \left(\left(\frac{8001}{4X^2} - \frac{247}{2} \right)^2 + \left(\frac{3387103.335}{X^3} - 247 \right)^2 \right) r_p + 0.00455677794 X^2$$

For 4.024470 < X < 23.935573 :

$$\Phi = \left(\frac{3387103.335}{X^3} - 247 \right)^2 r_p + 0.00455677794 X^2$$

For 23.935573 < X < 30.000000 :

$$\Phi = 0.00455677794 X^2$$

```
===============================
```

For r_p = 1.000000 :

r_p: 1.000000 , X_extr: 23.965000 , Number of iterations: 466

value of the non-penalized function at X = 23.965000 : 2.617054
value of the penalized function at X = 23.965000 : 2.617054

```
===============================
```

For r_p = 10.000000 :

r_p: 10.000000 , X_extr: 23.950000 , Number of iterations: 2

value of the non-penalized function at X = 23.950000 : 2.613779
value of the penalized function at X = 23.950000 : 2.613779

```
===============================
```

For r_p = 50.000000 :

r_p: 50.000000 , X_extr: 23.965000 , Number of iterations: 1

value of the non-penalized function at X = 23.965000 : 2.617054
value of the penalized function at X = 23.965000 : 2.617054

```
===============================
```

For r_p = 100.000000 :

r_p: 100.000000 , X_extr: 23.980000 , Number of iterations: 1

value of the non-penalized function at X = 23.980000 : 2.620331
value of the penalized function at X = 23.980000 : 2.620331

==============================

For r_p = 120.000000 :

r_p: 120.000000 , X_extr: 23.965000 , Number of iterations: 2

value of the non-penalized function at X = 23.965000 : 2.617054
value of the penalized function at X = 23.965000 : 2.617054

==============================

For r_p = 600.000000 :

r_p: 600.000000 , X_extr: 23.980000 , Number of iterations: 1

value of the non-penalized function at X = 23.980000 : 2.620331
value of the penalized function at X = 23.980000 : 2.620331

==============================

Module 6.10: Numerical determination of the minimum for the function (6.17) under consideration of the inequality constraints g_1 and g_2 in the range $0 \leq X \leq 30$ based on the exterior penalty function method (exact solution: $X_{extr} = 23.936$)

6.3   Problems from Chapter 4

4.5 Numerical Determination of the Minimum of an Unconstrained Function with Two Variables Based on Newton's Method

The following Listing 6.11 shows the entire wxMaxima code for the determination of the minimum of the objective function given in Eq. (4.27)

```
(% i12)    load("my_funs.mac")$

           fpprintprec:8$
           ratprint: false$

           c : 1.5$
           a : 1$
           b : 1$

           func(X) := -sqrt(c^2*(1-(X[1]^2/a^2)-(X[2]^2/b^2)))$

           eps : 1/1000$

           no_of_vars : 2$

           X_0 : [X[1]=-0.5,X[2]=0.5]$
           alpha_0 : 1$

           X_new : Newton_multi_variable_unconstrained(func,no_of_vars,X_0,alpha_0,
                              eps,"Kuhn_Tucker",true)$
i=1 X=[X[1]=-4.5474735*10^-13, X[2]=4.5474735*10^-13] func(X) =-1.5
Converged after 1 iterations!
```

Module 6.11: Numerical determination of the minimum for the function (4.27) based on the Newton's method for the initial point $X_0 = [-0.5\,0.5]^T$

6.4 Problems from Chapter 5

5.2 Numerical Determination of the Optimal Design of a Cantilever Beam: Constant Rectangular Cross-sectional Area

The objective function, i.e. the mass of the beam, can be stated as a function of the two design variables $X_1 = a$ and $X_2 = b$ as:

$$F(X_1, X_2) = m(X_1, X_2) = \varrho V = \varrho L X_1 X_2, \tag{6.20}$$

which is to be minimized under the following three inequality constraints (see Fig. 6.7), [1, 4]:

Fig. 6.7 Optimal design of a cantilever beam:
$L = 2540$ mm,
$F_0 = 2667$ N,
$E = 68948$ MPa,
$R_{p0.2} = 247$ MPa,
$\varrho = 2.691 \times 10^{-6}$ kg/mm³,
$r_1 = 0.03$ (original set of Eqs. (6.20)–(6.23)). Exact solution for the minimum:
$X_{1,extr} = 8.031$ mm,
$X_{2,extr} = 160.614$ mm
(indicated by the ● marker)

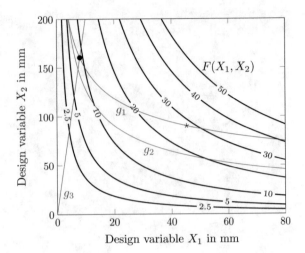

$$g_1(X_1, X_2) = \frac{4F_0 L^3}{E X_1 X_2^3} - r_1 L \leq 0 \quad \text{(displacement)}, \tag{6.21}$$

$$g_2(X_1, X_2) = \frac{6F_0 L}{X_1 X_2^2} - R_{p0.2} \leq 0 \quad \text{(normal stress)}, \tag{6.22}$$

$$g_3(X_1, X_2) = X_2 - 20 X_1 \leq 0 \quad \text{(height-to-width ratio)}. \tag{6.23}$$

Evaluating Fig. 6.7, it can be easily concluded that the minimum, under the given inequality constraints, is given by the intersection point of curves g_1 and g_3. A simple calculation reveals the *analytical* solution as:

$$X_{1,extr} = (20)^{-\frac{3}{4}} \times \left(\frac{4F_0 L^2}{E r_1} \right)^{\frac{1}{4}} = 8.031 \text{ mm}, \tag{6.24}$$

$$X_{2,extr} = 20 \times X_{1,extr} = 160.614 \text{ mm}. \tag{6.25}$$

The result from Problem 3.6, i.e. $a = 45.160$ and $2a = 90.3$ is indicated by a ⋆ marker in Fig. 6.7. For these geometrical dimensions, a total mass of $F(a, 2a) = m(a, 2a) = 27.9$ kg is obtained whereas the optimization for variable width and height gives $F(X_{1,extr}, X_{2,extr}) = m(X_{1,extr}, X_{2,extr}) = 8.8$ kg.

The following Listing 6.12 shows the entire wxMaxima code for the determination of the minimum of the objective function given in Eq. (6.20).

```
(% i12)    load("my_funs.mac")$

           fpprintprec:6$
           ratprint: false$
           eps : 1/1000

           L : 2540$
           E : 68948$
           ro : 2.691e-6$
           R_p02 : 247$
           F_0 : 2667$
           r_1 : 0.03$

           func_obj(X) := ro*L*X[1]*X[2]$

           g[1](X) :=(4*F_0*(L^3))/(E*X[1]*(X[2]^3)) - r_1*L$
           g[2](X) := (6*F_0*L)/(X[1]*(X[2]^2)) - R_p02$
           g[3](X) := X[2] - 20*X[1]$

           no_of_vars : 2$

           X_0 : [X[1]=1,X[2]=1]$
           alpha_0 : 1$
           r_p_0s : [0.5, 1, 10, 20, 50, 100]$

           gamma_value : 1$

           for r : 1 thru length(r_p_0s) do (
               print("==============="),
               print("For ", r_p = r_p_0s[r]),
               [X_new, pseudo_objective_fun_value] : Newton_multi_variable_constrained
                  (func_obj,no_of_vars,X_0,alpha_0,eps,r_p_0s[r], gamma_value,
                  "Kuhn_Tucker",false),
               print(X["1"] = rhs(X_new[1])),
               print(X["2"] = rhs(X_new[2])),
               printf(true, "The pseudo-objective function value at this point:
                      F = ~,6f", pseudo_objective_fun_value),
               X_0 : copy(X_new)
           )$
```

===============
For r_p=0.5
Converged after 18 iterations!
X[1]=8.02817
X[2]=160.591
The pseudo-objective function value at this point: F = 8.814248
===============
For r_p=1
Converged after 1 iterations!
X[1]=8.02944

X[2]=160.603
The pseudo-objective function value at this point: F = 8.815272
===============
For r_p=10
Converged after 2 iterations!
X[1]=8.03059
X[2]=160.613
The pseudo-objective function value at this point: F = 8.816194
===============
For r_p=20
Converged after 1 iterations!
X[1]=8.03065
X[2]=160.614
The pseudo-objective function value at this point: F = 8.816245
===============
For r_p=50
Converged after 1 iterations!
X[1]=8.03069
X[2]=160.614
The pseudo-objective function value at this point: F = 8.816276
===============
For r_p=100
Converged after 1 iterations!
X[1]=8.03071
X[2]=160.614
The pseudo-objective function value at this point: F = 8.816286

Module 6.12: Numerical determination of the minimum of the objective function $F(X_1, X_2)$ and the limitations due to three inequality constraints (see Eqs. (6.20)–(6.23))

5.3 Numerical Determination of the Optimal Design of a Simply Supported Beam: Constant Rectangular Cross-sectional Area

The objective function, i.e. the mass of the beam, can be stated as a function of the two design variables $X_1 = a$ and $X_2 = b$ as:

$$F(X_1, X_2) = m(X_1, X_2) = \varrho V = \varrho L X_1 X_2, \tag{6.26}$$

which is to be minimized under the following four inequality constraints (see Fig. 6.8), [1]:

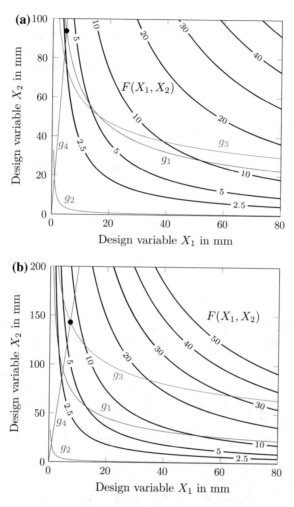

Fig. 6.8 Optimal design of a simply supported beam with $L = 2540\,\text{mm}$, $F_0 = 2667\,\text{N}$, $E = 68948\,\text{MPa}$, $R_{\text{p0.2}} = 247\,\text{MPa}$, $\varrho = 2.691 \times 10^{-6}\,\text{kg/mm}^3$ (original set of equations (6.26)–(6.30)): **a** $r_1 = 0.03$, exact solution for the minimum: $X_{1,\text{extr}} = 4.685\,\text{mm}$, $X_{2,\text{extr}} = 93.704\,\text{mm}$ (indicated by the • marker), **b** $r_1 = 0.003$, exact solution for the minimum: $X_{1,\text{extr}} = 7.140\,\text{mm}$, $X_{2,\text{extr}} = 142.809\,\text{mm}$ (indicated by the • marker)

$$g_1(X_1, X_2) = \frac{3F_0L}{2X_1X_2^2} - R_{\text{p0.2}} \leq 0 \quad \text{(normal stress)}, \tag{6.27}$$

$$g_2(X_1, X_2) = \frac{3F_0}{4X_1X_2} - \frac{R_{\text{p0.2}}}{2} \leq 0 \quad \text{(shear stress)}, \tag{6.28}$$

$$g_3(X_1, X_2) = \frac{F_0L^3}{4EX_1X_2^3} - r_1L \leq 0 \quad \text{(displacement)}, \tag{6.29}$$

$$g_4(X_1, X_2) = X_2 - 20X_1 \leq 0 \quad \text{(height-to-width ratio)}. \tag{6.30}$$

Evaluating Fig. 6.8, it can be easily concluded that the minimum, under the given inequality constraints, is given by the intersection point of curves g_1 and g_4 for the case $r_1 = 0.03$ and as the intersection point of curves g_3 and g_4 for the case $r_1 = 0.003$. A simple calculation reveals the *analytical* solutions as:

$$X_{1,\text{extr}}\Big|_{r_1=0.03} = (20)^{-\frac{2}{3}} \times \left(\frac{3F_0 L}{2R_{p0.2}}\right)^{\frac{1}{3}} = 4.685 \text{ mm}, \tag{6.31}$$

$$X_{2,\text{extr}}\Big|_{r_1=0.03} = 20 \times X_{1,\text{extr}} = 93.704 \text{ mm}, \tag{6.32}$$

or

$$X_{1,\text{extr}}\Big|_{r_1=0.003} = (20)^{-\frac{3}{4}} \times \left(\frac{F_0 L^2}{4Er_1}\right)^{\frac{1}{4}} = 7.140 \text{ mm}, \tag{6.33}$$

$$X_{2,\text{extr}}\Big|_{r_1=0.003} = 20 \times X_{1,\text{extr}} = 142.809 \text{ mm}. \tag{6.34}$$

The following Listing 6.13 shows the entire wxMaxima code for the determination of the minimum of the objective function given in Eq. (6.26) in the case of $r_1 = 0.03$.

```
(% i12)   load("my_funs.mac")$

          fpprintprec:6$
          ratprint: false$
          eps : 1/1000

          L : 2540$
          E : 68948$
          ro : 2.691e-6$
          R_p02 : 247$
          F_0 : 2667$
          r_1 : 0.03$

          func_obj(X) := ro*L*X[1]*X[2]$

          g[1](X) := (3*F_0*L)/(2*X[1]*(X[2]^2)) - R_p02$
          g[2](X) := (3*F_0)/(4*X[1]*(X[2])) - R_p02/2$
          g[3](X) := (F_0*(L^3))/(4*E*X[1]*(X[2]^3)) - r_1*L$
          g[4](X) := X[2]-20*X[1]$
          no_of_vars : 2$

          X_0 : [X[1]=1,X[2]=1]$
```

```
alpha_0 : 1$
r_p_0s : [0.5, 1, 10, 20, 50, 100]$

gamma_value : 1$

for r : 1 thru length(r_p_0s) do (
    print("=============="),
    print("For ", r_p = r_p_0s[r]),
    [X_new, pseudo_objective_fun_value] : Newton_multi_variable_constrained
        (func_obj,no_of_vars,X_0,alpha_0,eps,r_p_0s[r], gamma_value,
        "Kuhn_Tucker",false),
    print(X["1"] = rhs(X_new[1])),
    print(X["2"] = rhs(X_new[2])),
    printf(true, "The pseudo-objective function value at this point:
            F = ~,6f", pseudo_objective_fun_value),
    X_0 : copy(X_new)
)$
```

```
==============
For r_p=0.5
Converged after 37 iterations!
X[1]=4.68483
X[2]=93.7072
The pseudo-objective function value at this point: F = 3.000724
==============
For r_p=1
Converged after 1 iterations!
X[1]=4.68503
X[2]=93.7058
The pseudo-objective function value at this point: F = 3.000766
==============
For r_p=10
Converged after 1 iterations!
X[1]=4.6852
X[2]=93.7046
The pseudo-objective function value at this point: F = 3.000803
==============
For r_p=20
Converged after 1 iterations!
X[1]=4.68521
X[2]=93.7045
The pseudo-objective function value at this point: F = 3.000805
==============
For r_p=50
Converged after 1 iterations!
X[1]=4.68522
X[2]=93.7045
The pseudo-objective function value at this point: F = 3.000806
==============
For r_p=100
Converged after 1 iterations!
X[1]=4.68522
X[2]=93.7045
The pseudo-objective function value at this point: F = 3.000807
```

Module 6.13: Numerical determination of the minimum of the objective function $F(X_1, X_2)$ and the limitations due to four inequality constraints (see Eqs. (6.26)–(6.30)) in the case of $r_1 = 0.03$

5.4 Numerical Determination of the Optimal Design of a Short Cantilever Beam: Constant Rectangular Cross-sectional Area

The objective function, i.e. the mass of the beam, can be stated as a function of the two design variables $X_1 = a$ and $X_2 = b$ as:

$$F(X_1, X_2) = m(X_1, X_2) = \varrho V = \varrho L X_1 X_2, \tag{6.35}$$

which is to be minimized under the following three inequality constraints (see Fig. 6.9), [1]:

$$g_1(X_1, X_2) = \frac{6 F_0 L}{X_1 X_2^2} - R_{p0.2} \leq 0 \quad \text{(normal stress)}, \tag{6.36}$$

$$g_2(X_1, X_2) = \frac{3 F_0}{2 X_1 X_2} - \frac{R_{p0.2}}{2} \leq 0 \quad \text{(shear stress)}, \tag{6.37}$$

$$g_3(X_1, X_2) = X_2 - 20 X_1 \leq 0 \quad \text{(height-to-width ratio)}. \tag{6.38}$$

Evaluating Fig. 6.9, it can be easily concluded that the minimum, under the given inequality constraints, is given by the intersection point of curves g_1 and g_3. A simple calculation reveals the *analytical* solution as:

$$X_{1,\text{extr}} = 20^{-\frac{2}{3}} \times \left(\frac{6 F_0 L}{R_{p0.2}} \right)^{\frac{1}{3}} = 5.157 \text{ mm}, \tag{6.39}$$

$$X_{2,\text{extr}} = 20 \times X_{1,\text{extr}} = 103.135 \text{ mm}. \tag{6.40}$$

The result from Problem 3.8, i.e. $a = 23.936$ and $2a = 47.872$ is indicated by a $\star$ marker in Fig. 6.9. For these geometrical dimensions, a total mass of $F(a, 2a) = m(a, 2a) = 2.611$ kg is obtained whereas the optimization for variable width and height gives $F(X_{1,\text{extr}}, X_{2,\text{extr}}) = m(X_{1,\text{extr}}, X_{2,\text{extr}}) = 1.212$ kg.

The following Listing 6.14 shows the entire wxMaxima code for the determination of the minimum of the objective function given in Eq. (6.35).

```
(% i19)   load("my_funs.mac")$

          fpprintprec:6$
          ratprint: false$
          eps : 1/1000
          L : 846.33$
          E : 68948$
          ro : 2.691e-6$
          R_p02 : 247$
          F_0 : 2667$

          func_obj(X) := ro*L*X[1]*X[2]$

          g[1](X) := (6*F_0*L)/(X[1]*(X[2]^2)) - R_p02$
          g[2](X) := (3*F_0)/(2*X[1]*X[2]) - R_p02/2$
          g[3](X) := X[2] - 20*X[1]$

          no_of_vars : 2$

          X_0 : [X[1]=1,X[2]=1]$
          alpha_0 : 1$
          r_p_0s : [0.5, 1, 10, 20]$
          gamma_value : 1$

          for r : 1 thru length(r_p_0s) do (
              print("==============="),
              print("For ", r_p = r_p_0s[r]),
              [X_new, pseudo_objective_fun_value] : Newton_multi_variable_constrained
                 (func_obj,no_of_vars,X_0,alpha_0,eps,r_p_0s[r], gamma_value,
                  "Kuhn_Tucker",false),
              print(X["1"] = rhs(X_new[1])),
              print(X["2"] = rhs(X_new[2])),
              printf(true, "The pseudo-objective function value at this point:
                      F = ~,6f", pseudo_objective_fun_value),
              X_0 : copy(X_new)
          )$
```

===============
For r_p=0.5
Converged after 55 iterations!
X[1]=5.15592
X[2]=103.122
The pseudo-objective function value at this point: F = 1.210930
===============
For r_p=10
Converged after 1 iterations!
X[1]=5.15606
X[2]=103.121
The pseudo-objective function value at this point: F = 1.210936
===============
For r_p=20
Converged after 1 iterations!
X[1]=5.15607
X[2]=103.121
The pseudo-objective function value at this point: F = 1.210936

Module 6.14: Numerical determination of the minimum of the objective function $F(X_1, X_2)$ and the limitations due to three inequality constraints (see Eqs. (6.35)–(6.38))

Fig. 6.9 Optimal design of a
short cantilever beam:
$L = 846.67\,\text{mm}$,
$F_0 = 2667\,\text{N}$,
$E = 68948\,\text{MPa}$,
$R_{p0.2} = 247\,\text{MPa}$,
$\varrho = 2.691 \times 10^{-6}\,\text{kg/mm}^3$,
(original set of equations
(6.35)–(6.38)). Exact
solution for the minimum:
$X_{1,\text{extr}} = 5.157\,\text{mm}$,
$X_{2,\text{extr}} = 103.135\,\text{mm}$
(indicated by the • marker)

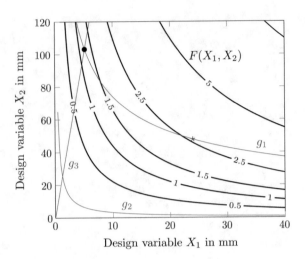

5.5 Optimization of a Stepped Cantilever Beam with Two Sections

The problem requires that the deformations as well as the stress distributions must
be calculated. This task can be achieved by a finite element 'hand calculation'. A
single stiffness matrix of an Euler-Bernoulli beam element can be stated as, see
[2, 3]:

$$K_i^{\text{e}} = \frac{E_i I_i}{L_i^3} \begin{bmatrix} 12 & -6L_i & -12 & -6L_i \\ -6L_i & 4L_i^2 & 6L_i & 2L_i^2 \\ -12 & 6L_i & 12 & 6L_i \\ -6L_i & 2L_i^2 & 6L_i & 4L_i^2 \end{bmatrix}. \tag{6.41}$$

Assembling the two elemental stiffness matrices K_i^{e} under the consideration of
$E_{\text{I}} = E_{\text{II}} = E$, $L_{\text{I}} = L_{\text{II}} = \frac{L}{2}$, $I_{\text{I}} = \frac{1}{12} a b_{\text{I}}^3$, and $I_{\text{II}} = \frac{1}{12} a b_{\text{II}}^3$ results in the following
global (reduced) system of equations:

$$E \begin{bmatrix} \left(\frac{8a\,b_{\text{II}}^3}{L^3} + \frac{8a\,b_{\text{I}}^3}{L^3} \right) & \left(\frac{2a\,b_{\text{I}}^3}{L^2} - \frac{2a\,b_{\text{II}}^3}{L^2} \right) & -\frac{8a\,b_{\text{II}}^3}{L^3} & -\frac{2a\,b_{\text{II}}^3}{L^2} \\ \left(\frac{2a\,b_{\text{I}}^3}{L^2} - \frac{2a\,b_{\text{II}}^3}{L^2} \right) & \left(\frac{2a\,b_{\text{II}}^3}{3L} + \frac{2a\,b_{\text{I}}^3}{3L} \right) & \frac{2a\,b_{\text{II}}^3}{L^2} & \frac{a\,b_{\text{II}}^3}{3L} \\ -\frac{8a\,b_{\text{II}}^3}{L^3} & \frac{2a\,b_{\text{II}}^3}{L^2} & \frac{8a\,b_{\text{II}}^3}{L^3} & \frac{2a\,b_{\text{II}}^3}{L^2} \\ -\frac{2a\,b_{\text{II}}^3}{L^2} & \frac{a\,b_{\text{II}}^3}{3L} & \frac{2a\,b_{\text{II}}^3}{L^2} & \frac{2a\,b_{\text{II}}^3}{3L} \end{bmatrix} \begin{bmatrix} u_{2z} \\ \varphi_{2y} \\ u_{3z} \\ \varphi_{3y} \end{bmatrix} = \begin{bmatrix} 0 \\ 0 \\ -F_0 \\ 0 \end{bmatrix}. \tag{6.42}$$

The last equation already considers that all degrees of freedom are zero at node 1. The
solution of the linear system of equations can be obtained, for example, by inverting
the global stiffness matrix and multiplying with the right-hand side, i.e. $u = K^{-1} f$,
to obtain the column matrix of nodal unknowns:

Table 6.6 Nodal values of the internal reactions bending moment (M_y) and shear force (Q_z) at each node

	Element I		Element II	
	Left node	Right node	Left node	Right node
M_y	$F_0 L$	$\frac{F_0 L}{2}$	$\frac{F_0 L}{2}$	0
Q_z	$-F_0$	$-F_0$	$-F_0$	$-F_0$

$$
\begin{bmatrix} u_{2z} \\ \varphi_{2y} \\ u_{3z} \\ \varphi_{3y} \end{bmatrix} = \begin{bmatrix} -\frac{5F_0 L^3}{4Ea\,b_{\mathrm{I}}^{3}} \\ \frac{9F_0 L^2}{2Ea\,b_{\mathrm{I}}^{3}} \\ -\frac{F_0 L^3\,(7b_{\mathrm{II}}^{3}+b_{\mathrm{I}}^{3})}{2Ea\,b_{\mathrm{I}}^{3}b_{\mathrm{II}}^{3}} \\ \frac{3F_0 L^2\,(3b_{\mathrm{II}}^{3}+b_{\mathrm{I}}^{3})}{2Ea\,b_{\mathrm{I}}^{3}b_{\mathrm{II}}^{3}} \end{bmatrix} .
\tag{6.43}
$$

The bending moment and shear force distributions within an element can be generally expressed as (index '1' refers to the start node while index '2' refers to the end node):

$$
M_y^e(x) = EI_y \left(\left[+\frac{6}{L^2} - \frac{12x}{L^3}\right] u_{1z} + \left[-\frac{4}{L} + \frac{6x}{L^2}\right] \varphi_{1y} \right.
$$
$$
\left. + \left[-\frac{6}{L^2} + \frac{12x}{L^3}\right] u_{2z} + \left[-\frac{2}{L} + \frac{6x}{L^2}\right] \varphi_{2y} \right) ,
\tag{6.44}
$$

$$
Q_z^e(x) = EI_y \left(\left[-\frac{12}{L^3}\right] u_{1z} + \left[+\frac{6}{L^2}\right] \varphi_{1y} + \left[+\frac{12}{L^3}\right] u_{2z} + \left[+\frac{6}{L^2}\right] \varphi_{2y} \right) .
\tag{6.45}
$$

Thus, the normal and shear stress distributions can be generally calculated as

$$
\sigma_x^e(x, z) = \frac{M_y^e(x)}{I_y} \times z ,
\tag{6.46}
$$

$$
\tau_{xz}^e(x, z) = \frac{Q_z^e(x)}{2I_y} \left[\left(\frac{b}{2}\right)^2 - z^2 \right] .
\tag{6.47}
$$

The nodal values of the internal reactions to calculate the normal and shear stresses are summarized in Table 6.6. It should be noted here that these nodal values based on the finite element approach are in this case equal to the analytical solution.

Fig. 6.10 Optimal design of a stepped cantilever beam with two sections: $L = 2540$ mm, $a = 45.16$ mm, $r_1 = 0.06$, $F_0 = 2667$ N, $E = 68948$ MPa, $R_{p0.2} = 247$ MPa, $\varrho = 2.691 \times 10^{-6}$ kg/mm^3, (original set of equations (6.52)–(6.59)). Exact solution for the minimum: $X_{1,\text{extr}} = 80.442$ mm, $X_{2,\text{extr}} = 49.455$ mm (indicated by the • marker)

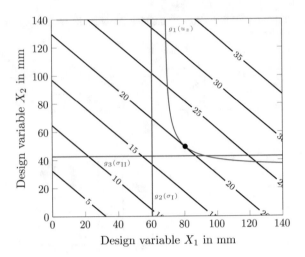

From Table 6.6 it can be concluded that the maximum normal stress in each element is attained at the left-hand node and that the shear stress is constant in each element. Thus, the critical stresses in each element can be stated as follow:

$$\sigma_{x,\text{I}} = \frac{6 F_0 L}{a b_{\text{I}}^2},$$ (6.48)

$$\sigma_{x,\text{II}} = \frac{3 F_0 L}{a b_{\text{II}}^2},$$ (6.49)

or for the shear stresses:

$$\tau_{xz,\text{I}} = -\frac{3 F_0}{2 a b_{\text{I}}},$$ (6.50)

$$\tau_{xz,\text{II}} = -\frac{3 F_0}{2 a b_{\text{II}}}.$$ (6.51)

The objective function, i.e. the mass of the stepped beam, can be stated as a function of the two design variables $b_{\text{I}} = X_1$ and $b_{\text{II}} = X_2$ as:

$$F(X_1, X_2) = \varrho L_{\text{I}} a X_1 + \varrho L_{\text{II}} a X_2 = \frac{\varrho L a}{2}(X_1 + X_2),$$ (6.52)

which is to be minimized under the following seven inequality constraints:

Fig. 6.11 Optimal design of a stepped cantilever beam with two sections: **a** magnification for smaller values of the design variables and **b** larger view

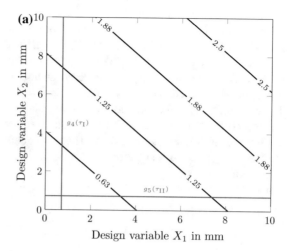

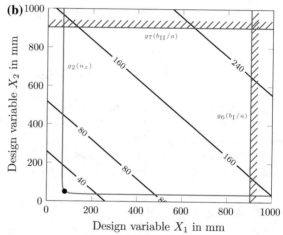

$$g_1(X_1, X_2) = +\frac{F_0 L^3 (X_1^3 + 7X_2^3)}{2EaX_1^3 X_2^3 r_1 L} - 1 \leq 0 \quad \text{(max. displ.)} , \tag{6.53}$$

$$g_2(X_1, X_2) = \frac{6F_0 L}{aX_1^2} - R_{p0.2} \leq 0 \quad \text{(normal stress in I)} , \tag{6.54}$$

$$g_3(X_1, X_2) = \frac{3F_0 L}{aX_2^2} - R_{p0.2} \leq 0 \quad \text{(normal stress in II)} , \tag{6.55}$$

$$g_4(X_1, X_2) = \frac{3F_0}{2aX_1} - \frac{R_{p0.2}}{2} \leq 0 \quad \text{(shear stress in I)} , \tag{6.56}$$

$$g_5(X_1, X_2) = \frac{3F_0}{2aX_2} - \frac{R_{p0.2}}{2} \leq 0 \quad \text{(shear stress in II)} , \tag{6.57}$$

$$g_6(X_1, X_2) = X_1 - 20a \leq 0 \quad \text{(height-to-width ratio in I)} , \tag{6.58}$$

$$g_7(X_1, X_2) = X_2 - 20a \leq 0 \quad \text{(height-to-width ratio in II)} . \tag{6.59}$$

The graphical representation of the objective function as well as the inequality constraints in the X_1-X_2 design space are given in Figs. 6.10 and 6.11.

It can be concluded from Figs. 6.10 and 6.11 that the consideration of g_1, g_2, and g_3 might be sufficient if a reasonable range of the design variables (and corresponding start values for the iterations) is considered. In addition, Fig. 6.10 indicates that the minimum is obtained for the case that g_1 is tangent to the objective function F. To derive an analytical expression for the minimum, it must be considered that the representations of the objective function F, in a representation $X_2(X_1) = \frac{2c}{\varrho La} - X_1$, are straight lines with a slope of -1. Thus, the condition for the minimum is that the function g_1 attains a slope of -1. Based on Eq. (6.54) we can reformulate the expression of g_1 as:

$$X_2 = \left(-\frac{F_0 L^3 X_1^3}{7F_0 L^3 - 2Ear_1 LX_1^3} \right)^{\frac{1}{3}} , \qquad (6.60)$$

from which we obtain, after a short calculation, the derivative as follows:

$$\left. \frac{dX_2}{dX_1} \right|_{g_1} = \frac{-7F_0^{\frac{4}{3}} L^{\frac{8}{3}}}{\left(7F_0 L^2 - 2Ear_1 X_1^3 \right)^{\frac{4}{3}}} \overset{!}{=} -1 . \qquad (6.61)$$

The last equation can be rearranged to have a zero on one side and the Newton's method can be used to find the root. Important is to search the root in a reasonable range, e.g. $70 \leq X_{1,\,extr} \leq 85$.

The following Listing 6.15 shows the entire wxMaxima code for the determination of the minimum of the objective function given in Eq. (6.52).

```
(% i39)  load("my_funs.mac")$

         fpprintprec:6$
         ratprint: false$
         eps : 1/1000$

         L : 2540$
         L1 : L/2$
         L2 : L/2$
         E : 68948$
         ro : 2.691e-6$
         R_p02 : 247$
         F_0 : 2667$
         r_1 : 0.06$
         a : 45.16$
```

```
l1 : (1/12)*a*X[1]^3$
l2 : (1/12)*a*X[2]^3$

u1z : 0$
phi1y : 0$
u2z : -5F_0*L^3/(4*E*a*X[1]^3)$
phi2y : 9*F_0*L^2/(2*E*a*X[1]^3)$
u3z : -F_0*L^3*(7*X[2]^3+X[1]^3)/(2*E*a*X[1]^3*X[2]^3)$
phi3y : 3*F_0*L^2*(3*X[2]^3+X[1]^3)/(2*E*a*X[1]^3*X[2]^3)$

func_obj(X) :=ro*L1*a*X[1] + ro*L2*a*X[2]$

/* Normal Stress - Sig(x,(b/2)) */
M_y : E*li*(((6/Li^2)-(12*x/Li^3))*uLz + ((-4/Li)+6*x/Li^2)*phiLy
    + ((-6/Li^2)+12*x/Li^3)*uRz + ((-2/Li)+6*x/Li^2)*phiRy)$
Sig : M_y*(b/2)/li$

/* Shear Stress - Tau(x,0) */
Q_z : E*li*((-12/Li^3)*uLz + (6/Li^2)*phiLy + (12/Li^3)*uRz + (6/Li^2)*phiRy)$
Tau : Q_z*(b/2)^2/(2*li)$

/* Element 1 */
g[1](X) := at(Sig - R_p02, [x=0, Li=L1, li=l1, uLz=u1z, phiLy=phi1y, uRz=u2z,
    phiRy=phi2y, b=X[1]])$
g[2](X) := at(Tau - R_p02, [x=0, Li=L1, li=l1, uLz=u1z, phiLy=phi1y,
    uRz=u2z, phiRy=phi2y, b=X[1]])$
g[3](X) := X[1] - 20*a$

/* Element 2 */
g[4](X) := at(Sig - R_p02, [x=0, Li=L2, li=l2, uLz=u2z, phiLy=phi2y, uRz=u3z,
    phiRy=phi3y, b=X[2]])$
g[5](X) := at(Tau - R_p02, [x=0, Li=L2, li=l2, uLz=u2z, phiLy=phi2y,
    uRz=u3z, phiRy=phi3y, b=X[2]])$
g[6](X) := X[2] - 20*a$
g[7](X) := -u3z - r_1*L$

no_of_vars : 2$

X_0 : [X[1]=50,X[2]=50]$
alpha_0 : 1$
r_p_0s : [0.05, 0.5, 1, 10, 20, 50, 100]$

gamma_value : 1$

for r : 1 thru length(r_p_0s) do (
    print("==============="),
    print("For ", r_p = r_p_0s[r]),
    [X_new, pseudo_objective_fun_value] :
    Newton_multi_variable_constrained(func_obj,no_of_vars,X_0,alpha_0,eps,
        r_p_0s[r], gamma_value,"Kuhn_Tucker",true),
    print(X["1"] = rhs(X_new[1])),
    print(X["2"] = rhs(X_new[2])),
    printf(true, "The pseudo-objective function value at this point:
    F = ~,6f", pseudo_objective_fun_value),
    X_0 : copy(X_new)
)$
```

```
==============
For r_p=0.05
i=1 X=[X[1]=71.5176,X[2]=72.2764] func(X)=22.2047
i=2 X=[X[1]=71.5464,X[2]=72.066] func(X)=22.1767
i=3 X=[X[1]=72.6876,X[2]=63.8444] func(X)=21.3813
i=4 X=[X[1]=74.8048,X[2]=57.2578] func(X)=20.658
i=5 X=[X[1]=78.4958,X[2]=51.3616] func(X)=20.0889
i=6 X=[X[1]=80.1703,X[2]=49.554] func(X)=20.0402
i=7 X=[X[1]=80.3184,X[2]=49.4539] func(X)=20.0384
i=8 X=[X[1]=80.3652,X[2]=49.4075] func(X)=20.0383
Converged after 8 iterations!
X[1]=80.3652
X[2]=49.4075
The pseudo-objective function value at this point: F = 20.038342
==============
For r_p=0.5
i=1 X=[X[1]=80.4343,X[2]=49.4499] func(X)=20.047
Converged after 1 iterations!
X[1]=80.4343
X[2]=49.4499
The pseudo-objective function value at this point: F = 20.046958
==============
For r_p=1
i=1 X=[X[1]=80.4381,X[2]=49.4523] func(X)=20.0474
Converged after 1 iterations!
X[1]=80.4381
X[2]=49.4523
The pseudo-objective function value at this point: F = 20.047439
==============
For r_p=10
i=1 X=[X[1]=80.4416,X[2]=49.4545] func(X)=20.0479
Converged after 1 iterations!
X[1]=80.4416
X[2]=49.4545
The pseudo-objective function value at this point: F = 20.047871
==============
For r_p=20
i=1 X=[X[1]=80.4417,X[2]=49.4546] func(X)=20.0479
Converged after 1 iterations!
X[1]=80.4417
X[2]=49.4546
The pseudo-objective function value at this point: F = 20.047895
==============
For r_p=50
i=1 X=[X[1]=80.4418,X[2]=49.4547] func(X)=20.0479
Converged after 1 iterations!
X[1]=80.4418
X[2]=49.4547
The pseudo-objective function value at this point: F = 20.047910
==============
For r_p=100
i=1 X=[X[1]=80.4419,X[2]=49.4547] func(X)=20.0479
Converged after 1 iterations!
X[1]=80.4419
X[2]=49.4547
The pseudo-objective function value at this point: F = 20.047914
```

Module 6.15: Numerical determination of the minimum of the objective function $F(X_1, X_2)$ and the limitations due to seven inequality constraints (see Eqs. (6.52)–(6.59))

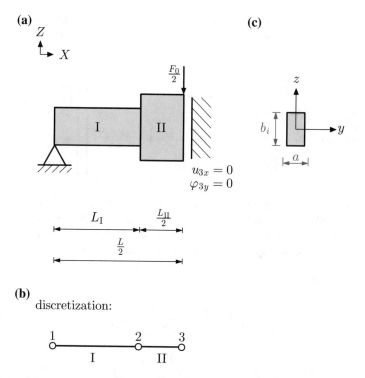

Fig. 6.12 Simply supported beam with three sections under consideration of symmetry: **a** general configuration, **b** discretization, and **c** cross-section of element i

5.6 Optimization of a Stepped Simply Supported Beam with Three Sections

The problem requires that the deformations as well as the stress distributions must be calculated. This task can be achieved by a finite element 'hand calculation'. A single stiffness matrix of an Euler-Bernoulli beam element can be stated as, see [2, 3]:

$$
K_i^e = \frac{E_i I_i}{L_i^3} \begin{bmatrix} 12 & -6L_i & -12 & -6L_i \\ -6L_i & 4L_i^2 & 6L_i & 2L_i^2 \\ -12 & 6L_i & 12 & 6L_i \\ -6L_i & 2L_i^2 & 6L_i & 4L_i^2 \end{bmatrix}. \tag{6.62}
$$

This particular problem allows to consider the symmetry in regard to the geometry and loading. Thus, the reduced system shown in Fig. 6.12 allows a quicker simulation than the entire structure.

Assembling the two elemental stiffness matrices K_i^e under the consideration of $E_I = E_{II} = E$, $L_I = L_{II} = \frac{L}{3}$, $I_I = \frac{1}{12}ab_I^3$, and $I_{II} = \frac{1}{12}ab_{II}^3$ results in the following global (reduced) system of equations:

$$E \begin{bmatrix} \frac{ab_I^3}{L} & \frac{9a\,b_I^3}{2L^2} & \frac{ab_I^3}{2L} & 0 \\ \frac{9a\,b_I^3}{2L^2} & \left(\frac{216a\,b_{II}^3}{L^3} + \frac{27a\,b_I^3}{L^3}\right) & \left(\frac{9a\,b_I^3}{2L^2} - \frac{18a\,b_{II}^3}{L^2}\right) & -\frac{216a\,b_{II}^3}{L^3} \\ \frac{ab_I^3}{2L} & \left(\frac{9a\,b_I^3}{2L^2} - \frac{18a\,b_{II}^3}{L^2}\right) & \left(\frac{2a\,b_{II}^3}{L} + \frac{ab_I^3}{L}\right) & \frac{18a\,b_{II}^3}{L^2} \\ 0 & -\frac{216a\,b_{II}^3}{L^3} & \frac{18a\,b_{II}^3}{L^2} & \frac{216a\,b_{II}^3}{L^3} \end{bmatrix} \begin{bmatrix} \varphi_{1y} \\ u_{2z} \\ \varphi_{2y} \\ u_{3z} \end{bmatrix} = \begin{bmatrix} 0 \\ 0 \\ 0 \\ \frac{-F_0}{2} \end{bmatrix}.$$

$$(6.63)$$

The last equation already considers that the translational degree of freedom is zero at node 1, whereas the rotational degree of freedom is zero at node 3. The solution of the linear system of equations can be obtained, for example, by inverting the global stiffness matrix and multiplying with the right-hand side, i.e. $u = K^{-1}f$, to obtain the column matrix of nodal unknowns:

$$\begin{bmatrix} \varphi_{1y} \\ u_{2z} \\ \varphi_{2y} \\ u_{3z} \end{bmatrix} = \begin{bmatrix} \frac{F_0 L^2 \left(4b_{II}^3 + 5b_I^3\right)}{12Ea\,b_I^3 b_{II}^3} \\ -\frac{F_0 L^3 \left(8b_{II}^3 + 15b_I^3\right)}{108Ea\,b_I^3 b_{II}^3} \\ \frac{5F_0 L^2}{12Ea\,b_{II}^3} \\ -\frac{F_0 L^3 \left(8b_{II}^3 + 19b_I^3\right)}{108Ea\,b_I^3 b_{II}^3} \end{bmatrix}.$$

$$(6.64)$$

The bending moment and shear force distributions within an element can be generally expressed as (index '1' refers to the start node while index '2' refers to the end node):

$$M_y^e(x) = EI_y \left(\left[+\frac{6}{L^2} - \frac{12x}{L^3} \right] u_{1z} + \left[-\frac{4}{L} + \frac{6x}{L^2} \right] \varphi_{1y} \right. \\ \left. + \left[-\frac{6}{L^2} + \frac{12x}{L^3} \right] u_{2z} + \left[-\frac{2}{L} + \frac{6x}{L^2} \right] \varphi_{2y} \right),$$

$$(6.65)$$

$$Q_z^e(x) = EI_y \left(\left[-\frac{12}{L^3} \right] u_{1z} + \left[+\frac{6}{L^2} \right] \varphi_{1y} + \left[+\frac{12}{L^3} \right] u_{2z} + \left[+\frac{6}{L^2} \right] \varphi_{2y} \right). \quad (6.66)$$

Thus, the normal and shear stress distributions can be generally calculated as

Table 6.7 Nodal values of the internal reactions bending moment (M_y) and shear force (Q_z) at each node

	Element I		Element II	
	Left node	Right node	Left node	Right node
M_y	0	$-\frac{F_0 L}{6}$	$-\frac{F_0 L}{6}$	$-\frac{F_0 L}{4}$
Q_z	$-\frac{F_0}{2}$	$-\frac{F_0}{2}$	$-\frac{F_0}{2}$	$-\frac{F_0}{2}$

$$\sigma_x^e(x, z) = \frac{M_y^e(x)}{I_y} \times z, \tag{6.67}$$

$$\tau_{xz}^e(x, z) = \frac{Q_z^e(x)}{2 I_y} \left[\left(\frac{b}{2} \right)^2 - z^2 \right]. \tag{6.68}$$

The nodal values of the internal reactions to calculate the normal and shear stresses are summarized in Table 6.7. It should be noted here that these nodal values based on the finite element approach are in this case equal to the analytical solution. From Table 6.7 it can be concluded that the maximum normal stress in each element is attained at the right-hand node and that the shear stress is constant in each element. Thus, the critical stresses in each element can be stated as follow:

$$\sigma_{x,\mathrm{I}} = -\frac{F_0 L}{a\, b_{\mathrm{I}}^2}, \tag{6.69}$$

$$\sigma_{x,\mathrm{II}} = -\frac{3 F_0 L}{2a\, b_{\mathrm{II}}^2}, \tag{6.70}$$

or for the shear stresses:

$$\tau_{xz,\mathrm{I}} = -\frac{3 F_0}{4a\, b_{\mathrm{I}}}, \tag{6.71}$$

$$\tau_{xz,\mathrm{II}} = -\frac{3 F_0}{4a\, b_{\mathrm{II}}}. \tag{6.72}$$

The objective function, i.e. the mass of the stepped beam, can be stated as a function of the two design variables $b_{\mathrm{I}} = X_1$ and $b_{\mathrm{II}} = X_2$ as:

$$F(X_1, X_2) = \varrho L_{\mathrm{I}} a X_1 + \varrho L_{\mathrm{II}} a X_2 = \varrho L a \left(\frac{X_1}{3} + \frac{X_2}{6} \right), \tag{6.73}$$

which is to be minimized under the following seven inequality constraints:

$$g_1(X_1, X_2) = +\frac{F_0 L^3 (19X_1^3 + 8X_2^3)}{108 E a X_1^3 X_2^3 r_1 L} - 1 \leq 0 \quad \text{(max. displ.)}, \qquad (6.74)$$

$$g_2(X_1, X_2) = \frac{F_0 L}{a X_1^2} - R_{p0.2} \leq 0 \quad \text{(normal stress in I)}, \qquad (6.75)$$

$$g_3(X_1, X_2) = \frac{3 F_0 L}{2 a X_2^2} - R_{p0.2} \leq 0 \quad \text{(normal stress in II)}, \qquad (6.76)$$

$$g_4(X_1, X_2) = \frac{3 F_0}{4 a X_1} - \frac{R_{p0.2}}{2} \leq 0 \quad \text{(shear stress in I)}, \qquad (6.77)$$

$$g_5(X_1, X_2) = \frac{3 F_0}{4 a X_2} - \frac{R_{p0.2}}{2} \leq 0 \quad \text{(shear stress in II)}, \qquad (6.78)$$

$$g_6(X_1, X_2) = X_1 - 20a \leq 0 \quad \text{(height-to-width ratio in I)}, \qquad (6.79)$$

$$g_7(X_1, X_2) = X_2 - 20a \leq 0 \quad \text{(height-to-width ratio in II)}. \qquad (6.80)$$

The graphical representation of the objective function as well as the inequality constraints (g_1, g_2, g_3) in the X_1–X_2 design space are given in Fig. 6.13.

It can be concluded from Fig. 6.13 that the consideration of g_1, g_2, and g_3 might be sufficient if a reasonable range of the design variables (and corresponding start values for the iterations) is considered. In addition, Fig. 6.13 indicates that the minimum is obtained for the case that g_1 is tangent to the objective function F. To derive an analytical expression for the minimum, it must be considered that the representations of the objective function F, in a representation $X_2(X_1) = \frac{6c}{\varrho L a} - 2X_1$, are straight lines with a slope of -2. Thus, the condition for the minimum is that the function g_1 attains a slope of -2. Based on Eq. (6.74) we can reformulate the expression of g_1 as:

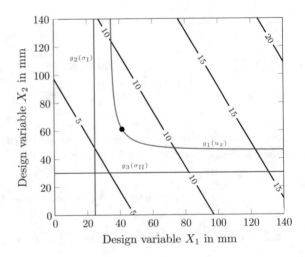

Fig. 6.13 Optimal design of a simply supported beam with three sections (consideration of symmetry): $L = 2540$ mm, $a = 45.16$ mm, $r_1 = 0.01$, $F_0 = 2667$ N, $E = 68948$ MPa, $R_{p0.2} = 247$ MPa, $\varrho = 2.691 \times 10^{-6}$ kg/mm³, (original set of equations (6.73)–(6.80)). Exact solution for the minimum: $X_{1,\text{extr}} = 41.437$ mm, $X_{2,\text{extr}} = 61.173$ mm (indicated by the ● marker)

$$X_2 = \left(-\frac{19 F_0 L^3 X_1^3}{8 F_0 L^3 - 108 E a r_1 L X_1^3}\right)^{\frac{1}{3}}, \tag{6.81}$$

from which we obtain, after a short calculation, the derivative as follows:

$$\left.\frac{dX_2}{dX_1}\right|_{g_1} = \frac{219^{\frac{1}{3}} F^{\frac{4}{3}} L^{\frac{8}{3}}}{\left(8F\,L^2 - 108 E a\,r_1 X_1^{\,3}\right)^{\frac{1}{3}} \left(27 E a\,r_1 X_1^{\,3} - 2 F\,L^2\right)} \overset{!}{=} -2. \tag{6.82}$$

The last equation can be rearranged to have a zero on one side and the Newton's method can be used to find the root. Important is to search the root in a reasonable range, e.g. $40 \le X_{1,\,\text{extr}} \le 60$.

The following Listing 6.16 shows the entire wxMaxima code for the determination of the minimum of the objective function given in Eq. (6.73).

```
(% i39)   load("my_funs.mac")$

          fpprintprec:6$
          ratprint: false$
          eps : 1/1000$

          L : 2540$
          L1 : L/3$
          L2 : L/3$
          E : 68948$
          ro : 2.691e-6$
          R_p02 : 247$
          F_0 : 2667$
          r_1 : 0.01$
          a : 45.16$

          I1 : (1/12)*a*X[1]^3$
          I2 : (1/12)*a*X[2]^3$

          u1z : 0$
          phi1y : F_0*L^2*(4*X[2]^3+5*X[1]^3)/(12*E*a*X[1]^3*X[2]^3)$
          u2z : -F_0*L^3*(8*X[2]^3+15*X[1]^3)/(108*E*a*X[1]^3*X[2]^3)$
          phi2y : 5*F_0*L^2/(12*E*a*X[2]^3)$
          u3z : -F_0*L^3*(8*X[2]^3+19*X[1]^3)/(108*E*a*X[1]^3*X[2]^3)$
          phi3y : 0$

          func_obj(X) := ro*L1*a*X[1] + ro*L2*a*X[2]$

          /* Normal Stress - Sig(x,(b/2)) */
          M_y : E*Ii*(((6/Li^2)-(12*x/Li^3))*uLz + ((-4/Li)+6*x/Li^2)*phiLy +
```

```
        ((-6/Li^2)+12*x/Li^3)*uRz + ((-2/Li)+6*x/Li^2)*phiRy)$
Sig : M_y*(b/2)/Ii$

/* Shear Stress - Tau(x,0) */
Q_z : E*Ii*((-12/Li^3)*uLz + (6/Li^2)*phiLy + (12/Li^3)*uRz + (6/Li^2)*phiRy)$
Tau : Q_z*(b/2)^2/(2*Ii)$

/* Element 1 */
g[1](X) := at(Sig - R_p02, [x=L1, Li=L1, Ii=I1, uLz=u1z, phiLy=phi1y,
     uRz=u2z, phiRy=phi2y, b=X[1]])$
g[2](X) := at(Tau - R_p02, [x=L1, Li=L1, Ii=I1, uLz=u1z, phiLy=phi1y,
     uRz=u2z, phiRy=phi2y, b=X[1]])$
g[3](X) := X[1] - 20*a$

/* Element 2 */
g[4](X) := at(Sig - R_p02, [x=L2, Li=L2, Ii=I2, uLz=u2z, phiLy=phi2y,
     uRz=u3z, phiRy=phi3y, b=X[2]])$
g[5](X) := at(Tau - R_p02, [x=L2, Li=L2, Ii=I2, uLz=u2z, phiLy=phi2y,
     uRz=u3z, phiRy=phi3y, b=X[2]])$
g[6](X) := X[2] - 20*a$
g[7](X) := -u3z - r_1*L$

/* Solution */
no_of_vars : 2$

X_0 : [X[1]=50,X[2]=50]$
alpha_0 : 1$
gamma_value : 1$
r_p_0s : [0.05, 0.5, 1, 10, 20, 50, 100]$

for r : 1 thru length(r_p_0s) do (
   print("=============="),
   print("For ", r_p = r_p_0s[r]),
   [X_new, pseudo_objective_fun_value] :
   Newton_multi_variable_constrained(func_obj,no_of_vars,X_0,alpha_0,eps,
      r_p_0s[r], gamma_value,"Kuhn_Tucker",true),
   print(X["1"] = rhs(X_new[1])),
   print(X["2"] = rhs(X_new[2])),
   printf(true, "The pseudo-objective function value at this point:
    F = ~,6f", pseudo_objective_fun_value),
   X_0 : copy(X_new)
)$
```

```
===============
For r_p=0.05
i=1 X=[X[1]=43.4635,X[2]=54.6909] func(X)=7.56368
i=2 X=[X[1]=42.3068,X[2]=58.5801] func(X)=7.38557
i=3 X=[X[1]=40.9799,X[2]=60.3152] func(X)=7.36569
i=4 X=[X[1]=40.9461,X[2]=60.4447] func(X)=7.36559
Converged after 4 iterations!
X[1]=40.9461
X[2]=60.4447
The pseudo-objective function value at this point: F = 7.365593
===============
For r_p=0.5
i=1 X=[X[1]=41.3855,X[2]=61.0938] func(X)=7.40594
Converged after 1 iterations!
X[1]=41.3855
X[2]=61.0938
The pseudo-objective function value at this point: F = 7.405941
===============
For r_p=1
i=1 X=[X[1]=41.4113,X[2]=61.1335] func(X)=7.40829
Converged after 1 iterations!
X[1]=41.4113
X[2]=61.1335
The pseudo-objective function value at this point: F = 7.408288
===============
For r_p=10
i=1 X=[X[1]=41.4349,X[2]=61.1686] func(X)=7.41041
Converged after 1 iterations!
X[1]=41.4349
X[2]=61.1686
The pseudo-objective function value at this point: F = 7.410410
===============
For r_p=20
i=1 X=[X[1]=41.436,X[2]=61.171] func(X)=7.41053
Converged after 1 iterations!
X[1]=41.436
X[2]=61.171
The pseudo-objective function value at this point: F = 7.410528
===============
For r_p=50
i=1 X=[X[1]=41.4367,X[2]=61.1724] func(X)=7.4106
Converged after 1 iterations!
X[1]=41.4367
X[2]=61.1724
The pseudo-objective function value at this point: F = 7.410599
===============
For r_p=100
i=1 X=[X[1]=41.4369,X[2]=61.1729] func(X)=7.41062
Converged after 1 iterations!
X[1]=41.4369
X[2]=61.1729
The pseudo-objective function value at this point: F = 7.410623
```

Module 6.16: Numerical determination of the minimum of the objective function $F(X_1, X_2)$ and the limitations due to seven inequality constraints (see Eqs. (6.73)–(6.80))

5.7 Optimization of a Stepped Cantilever Beam with Three Sections

A single stiffness matrix of an Euler-Bernoulli beam element can be stated as, see [2, 3]:

$$
\boldsymbol{K}_i^e = \frac{E_i I_i}{L_i^3}
\begin{bmatrix}
12 & -6L_i & -12 & -6L_i \\
-6L_i & 4L_i^2 & 6L_i & 2L_i^2 \\
-12 & 6L_i & 12 & 6L_i \\
-6L_i & 2L_i^2 & 6L_i & 4L_i^2
\end{bmatrix}.
\tag{6.83}
$$

Assembling the three elemental stiffness matrices $\boldsymbol{K}_i^e$ under the consideration of $E_{\mathrm{I}} = E_{\mathrm{II}} = E_{\mathrm{III}} = E$ and $L_{\mathrm{I}} = L_{\mathrm{II}} = L_{\mathrm{III}} = \frac{L}{3}$ results in the following global stiffness matrix:

$$
\boldsymbol{K} =
$$

$$
E
\begin{bmatrix}
\left(\frac{324 I_{\mathrm{II}}}{L^3} + \frac{324 I_{\mathrm{I}}}{L^3}\right) & \left(\frac{54 I_{\mathrm{I}}}{L^2} - \frac{54 I_{\mathrm{II}}}{L^2}\right) & -\frac{324 I_{\mathrm{II}}}{L^3} & -\frac{54 I_{\mathrm{II}}}{L^2} & 0 & 0 \\
\left(\frac{54 I_{\mathrm{I}}}{L^2} - \frac{54 I_{\mathrm{II}}}{L^2}\right) & \left(\frac{12 I_{\mathrm{II}}}{L} + \frac{12 I_{\mathrm{I}}}{L}\right) & \frac{54 I_{\mathrm{II}}}{L^2} & \frac{6 I_{\mathrm{II}}}{L} & 0 & 0 \\
-\frac{324 I_{\mathrm{II}}}{L^3} & \frac{54 I_{\mathrm{II}}}{L^2} & \left(\frac{324 I_{\mathrm{III}}}{L^3} + \frac{324 I_{\mathrm{II}}}{L^3}\right) & \left(\frac{54 I_{\mathrm{II}}}{L^2} - \frac{54 I_{\mathrm{III}}}{L^2}\right) & -\frac{324 I_{\mathrm{III}}}{L^3} & -\frac{54 I_{\mathrm{III}}}{L^2} \\
-\frac{54 I_{\mathrm{II}}}{L^2} & \frac{6 I_{\mathrm{II}}}{L} & \left(\frac{54 I_{\mathrm{II}}}{L^2} - \frac{54 I_{\mathrm{III}}}{L^2}\right) & \left(\frac{12 I_{\mathrm{III}}}{L} + \frac{12 I_{\mathrm{II}}}{L}\right) & \frac{54 I_{\mathrm{III}}}{L^2} & \frac{6 I_{\mathrm{III}}}{L} \\
0 & 0 & -\frac{324 I_{\mathrm{III}}}{L^3} & \frac{54 I_{\mathrm{III}}}{L^2} & \frac{324 I_{\mathrm{III}}}{L^3} & \frac{54 I_{\mathrm{III}}}{L^2} \\
0 & 0 & -\frac{54 I_{\mathrm{III}}}{L^2} & \frac{6 I_{\mathrm{III}}}{L} & \frac{54 I_{\mathrm{III}}}{L^2} & \frac{12 I_{\mathrm{III}}}{L}
\end{bmatrix}.
$$

$$\tag{6.84}$$

The last equation already considers that all degrees of freedom are zero at node 1. The solution of the linear system of equations can be obtained, for example, by inverting the global stiffness matrix and multiplying with the right-hand side, i.e. $\boldsymbol{u} = \boldsymbol{K}^{-1} \boldsymbol{f}$, to obtain the column matrix of nodal unknowns:

$$
\begin{bmatrix}
u_{2Z} \\
\varphi_{2Y} \\
u_{3Z} \\
\varphi_{3Y} \\
u_{4Z} \\
\varphi_{4Y}
\end{bmatrix}
=
\begin{bmatrix}
-\frac{4 F_0 L^3}{81 E\, I_{\mathrm{I}}} \\
\frac{5 F_0 L^2}{18 E\, I_{\mathrm{I}}} \\
-\frac{F_0 (23 I_{\mathrm{II}} + 5 I_{\mathrm{I}}) L^3}{162 E\, I_{\mathrm{I}} I_{\mathrm{II}}} \\
\frac{F_0 (5 I_{\mathrm{II}} + 3 I_{\mathrm{I}}) L^2}{18 E\, I_{\mathrm{I}} I_{\mathrm{II}}} \\
-\frac{F_0 (19 I_{\mathrm{II}} I_{\mathrm{III}} + 7 I_{\mathrm{I}} I_{\mathrm{III}} + I_{\mathrm{I}} I_{\mathrm{II}}) L^3}{81 E\, I_{\mathrm{I}} I_{\mathrm{II}} I_{\mathrm{III}}} \\
\frac{F_0 (5 I_{\mathrm{II}} I_{\mathrm{III}} + 3 I_{\mathrm{I}} I_{\mathrm{III}} + I_{\mathrm{I}} I_{\mathrm{II}}) L^2}{18 E\, I_{\mathrm{I}} I_{\mathrm{II}} I_{\mathrm{III}}}
\end{bmatrix}.
\tag{6.85}
$$

The bending moment and shear force distributions within an element can be generally expressed as (index '1' refers to the start node while index '2' refers to the end node):

Table 6.8 Nodal values of the internal reactions bending moment (M_y) and shear force (Q_z) at each node

	Element I		Element II		Element III	
	Left node	Right node	Left node	Right node	Left node	Right node
M_y	$F_0 L$	$\frac{2F_0 L}{3}$	$\frac{2F_0 L}{3}$	$-\frac{F_0 L}{3}$	$-\frac{F_0 L}{3}$	0
Q_z	$-F_0$	$-F_0$	$-F_0$	$-F_0$	$-F_0$	$-F_0$

$$M_y^e(x) = EI_y \left(\left[+\frac{6}{L^2} - \frac{12x}{L^3} \right] u_{1z} + \left[-\frac{4}{L} + \frac{6x}{L^2} \right] \varphi_{1y} \right.$$
$$\left. + \left[-\frac{6}{L^2} + \frac{12x}{L^3} \right] u_{2z} + \left[-\frac{2}{L} + \frac{6x}{L^2} \right] \varphi_{2y} \right) , \quad (6.86)$$

$$Q_z^e(x) = EI_y \left(\left[-\frac{12}{L^3} \right] u_{1z} + \left[+\frac{6}{L^2} \right] \varphi_{1y} + \left[+\frac{12}{L^3} \right] u_{2z} + \left[+\frac{6}{L^2} \right] \varphi_{2y} \right) . \quad (6.87)$$

The nodal values of the internal reactions to calculate the normal and shear stresses are summarized in Table 6.8. It should be noted here that these nodal values based on the finite element approach are in this case equal to the analytical solution.

From Table 6.7 it can be concluded that the maximum bending moment in each element is attained at the left-hand node and that the shear forces is constant in each element.

The normal and shear stress distributions can be generally calculated as

$$\sigma_x^e(x, z) = \frac{M_y^e(x)}{I_y} \times z , \quad (6.88)$$

$$\tau_{xz}^e(x, z) = \frac{Q_z^e(x)}{2I_y} \left[\left(\frac{b}{2} \right)^2 - z^2 \right] , \quad (6.89)$$

or as summarized in Tables 6.9, 6.10 and 6.11 for all three elements.

Thus, the maximum normal stress in each element is attained at the left-hand node and that the shear stress is constant in each element. Thus, the critical stresses in each element can be stated as follow:

$$\sigma_{x,\text{I}} = \frac{6F_0 L}{a_\text{I} b_\text{I}^2} , \quad (6.90)$$

$$\sigma_{x,\text{II}} = \frac{4F_0 L}{a_\text{II} b_\text{II}^2} , \quad (6.91)$$

$$\sigma_{x,\text{III}} = \frac{2F_0 L}{a_\text{III} b_\text{III}^2} , \quad (6.92)$$

Table 6.9 Maximal normal and shear stress values at the nodes of element I

Node	Stress value
LHS	$\sigma_x^{e,I}(0, b_I/2) = E\left(\left[+\dfrac{6}{L_I^2} - 0\right]0 + \left[-\dfrac{4}{L_I} + 0\right]0\right.$ $\left. + \left[-\dfrac{6}{L_I^2} + 0\right]u_{2z} + \left[-\dfrac{2}{L_I} + 0\right]\varphi_{2y}\right) \times \dfrac{b_I}{2}$ $\tau_{xz}^{e,I}(0,0) = \dfrac{E}{2}\left(\left[-\dfrac{12}{L_I^3}\right]0 + \left[+\dfrac{6}{L_I^2}\right]0 + \left[+\dfrac{12}{L_I^3}\right]u_{2z} + \left[+\dfrac{6}{L_I^2}\right]\varphi_{2y}\right) \times \left(\dfrac{b_I}{2}\right)^2$
RHS	$\sigma_x^{e,I}(L_I, b_I/2) = E\left(\left[+\dfrac{6}{L_I^2} - \dfrac{12 L_I}{L_I^3}\right]0 + \left[-\dfrac{4}{L_I} + \dfrac{6 L_I}{L_I^2}\right]0\right.$ $\left. + \left[-\dfrac{6}{L_I^2} + \dfrac{12 L_I}{L_I^3}\right]u_{2z} + \left[-\dfrac{2}{L_I} + \dfrac{6 L_I}{L_I^2}\right]\varphi_{2y}\right) \times \dfrac{b_I}{2}$ $\tau_{xz}^{e,I}(L_I, 0) = \dfrac{E}{2}\left(\left[-\dfrac{12}{L_I^3}\right]0 + \left[+\dfrac{6}{L_I^2}\right]0 + \left[+\dfrac{12}{L_I^3}\right]u_{2z} + \left[+\dfrac{6}{L_I^2}\right]\varphi_{2y}\right) \times \left(\dfrac{b_I}{2}\right)^2$

Table 6.10 Maximal normal and shear stress values at the nodes of element II

Node	Stress value
LHS	$\sigma_x^{e,II}(0, b_{II}/2) = E\left(\left[+\dfrac{6}{L_{II}^2} - 0\right]u_{2z} + \left[-\dfrac{4}{L_{II}} + 0\right]\varphi_{2y}\right.$ $\left. + \left[-\dfrac{6}{L_{II}^2} + 0\right]u_{3z} + \left[-\dfrac{2}{L_{II}} + 0\right]\varphi_{3y}\right) \times \dfrac{b_{II}}{2}$ $\tau_{xz}^{e,II}(0,0) =$ $\dfrac{E}{2}\left(\left[-\dfrac{12}{L_{II}^3}\right]u_{2z} + \left[+\dfrac{6}{L_{II}^2}\right]\varphi_{2y} + \left[+\dfrac{12}{L_{II}^3}\right]u_{3z} + \left[+\dfrac{6}{L_{II}^2}\right]\varphi_{3y}\right) \times \left(\dfrac{b_{II}}{2}\right)^2$
RHS	$\sigma_x^{e,II}(L_{II}, b_{II}/2) = E\left(\left[+\dfrac{6}{L_{II}^2} - \dfrac{12 L_{II}}{L_{II}^3}\right]u_{2z} + \left[-\dfrac{4}{L_{II}} + \dfrac{6 L_{II}}{L_{II}^2}\right]\varphi_{2y}\right.$ $\left. + \left[-\dfrac{6}{L_{II}^2} + \dfrac{12 L_{II}}{L_{II}^3}\right]u_{3z} + \left[-\dfrac{2}{L_{II}} + \dfrac{6 L_{II}}{L_{II}^2}\right]\varphi_{3y}\right) \times \dfrac{b_{II}}{2}$ $\tau_{xz}^{e,II}(L_{II}, 0) =$ $\dfrac{E}{2}\left(\left[-\dfrac{12}{L_{II}^3}\right]u_{2z} + \left[+\dfrac{6}{L_{II}^2}\right]\varphi_{2y} + \left[+\dfrac{12}{L_{II}^3}\right]u_{3z} + \left[+\dfrac{6}{L_{II}^2}\right]\varphi_{3y}\right) \times \left(\dfrac{b_{II}}{2}\right)^2$

Table 6.11 Maximal normal and shear stress values at the nodes of element III

Node	Stress value
LHS	$\sigma_x^{e,III}(0, b_{III}/2) = E\left(\left[+\dfrac{6}{L_{III}^2} - 0\right]u_{3z} + \left[-\dfrac{4}{L_{III}} + 0\right]\varphi_{3y}\right.$ $\left. + \left[-\dfrac{6}{L_{III}^2} + 0\right]u_{4z} + \left[-\dfrac{2}{L_{III}} + 0\right]\varphi_{4y}\right) \times \dfrac{b_{III}}{2}$ $\tau_{xz}^{e,III}(0, 0) =$ $\dfrac{E}{2}\left(\left[-\dfrac{12}{L_{III}^3}\right]u_{3z} + \left[+\dfrac{6}{L_{III}^2}\right]\varphi_{3y} + \left[+\dfrac{12}{L_{III}^3}\right]u_{4z} + \left[+\dfrac{6}{L_{III}^2}\right]\varphi_{4y}\right) \times \left(\dfrac{b_{III}}{2}\right)^2$
RHS	$\sigma_x^{e,III}(L_{III}, b_{III}/2) = E\left(\left[+\dfrac{6}{L_{III}^2} - \dfrac{12L_{III}}{L_{III}^3}\right]u_{3z} + \left[-\dfrac{4}{L_{III}} + \dfrac{6L_{III}}{L_{III}^2}\right]\varphi_{3y}\right.$ $\left. + \left[-\dfrac{6}{L_{III}^2} + \dfrac{12L_{III}}{L_{III}^3}\right]u_{4z} + \left[-\dfrac{2}{L_{III}} + \dfrac{6L_{III}}{L_{III}^2}\right]\varphi_{4y}\right) \times \dfrac{b_{III}}{2}$ $\tau_{xz}^{e,III}(L_{III}, 0) =$ $\dfrac{E}{2}\left(\left[-\dfrac{12}{L_{III}^3}\right]u_{3z} + \left[+\dfrac{6}{L_{III}^2}\right]\varphi_{3y} + \left[+\dfrac{12}{L_{III}^3}\right]u_{4z} + \left[+\dfrac{6}{L_{III}^2}\right]\varphi_{4y}\right) \times \left(\dfrac{b_{III}}{2}\right)^2$

or for the shear stresses:

$$\tau_{xz,I} = -\frac{3F_0}{2a_I b_I}, \tag{6.93}$$

$$\tau_{xz,II} = -\frac{3F_0}{2a_{II} b_{II}}, \tag{6.94}$$

$$\tau_{xz,III} = -\frac{3F_0}{2a_{III} b_{III}}. \tag{6.95}$$

The objective function, i.e. the mass of the stepped beam, can be stated as a function of the six design variables $a_I = X_1$, $b_I = X_2$, $a_{II} = X_3$, $b_{II} = X_4$, and $a_{III} = X_5$, $b_{III} = X_6$, as:

$$F(X_1, \ldots, X_6) = \varrho L_I X_1 X_2 + \varrho L_{II} X_3 X_4 + \varrho L_{III} X_5 X_6 \tag{6.96}$$

$$= \frac{\varrho L}{3}(X_1 X_2 + X_3 X_4 + X_5 X_6), \tag{6.97}$$

Table 6.12 Nodal values of the internal reactions bending moment (M_y) and shear force (Q_z) at each node

	Element I		Element II		Element III		Element IV	
	Left node	Right node	Left node	Right node	Left node	Right node	Left node	Right node
M_y	0	$-\frac{F_0 L}{8}$	$-\frac{F_0 L}{8}$	$-\frac{F_0 L}{4}$	$-\frac{F_0 L}{4}$	$-\frac{F_0 L}{8}$	$-\frac{F_0 L}{8}$	0
Q_z	$-\frac{F_0}{2}$	$-\frac{F_0}{2}$	$-\frac{F_0}{2}$	$-\frac{F_0}{2}$	$\frac{F_0}{2}$	$\frac{F_0}{2}$	$\frac{F_0}{2}$	$\frac{F_0}{2}$

which is to be minimized under the following 10 inequality constraints:

$$g_1 = \frac{F_0 \left(19 I_{II} I_{III} + 7 I_I I_{III} + I_I I_{II}\right) L^3}{81 E \, I_I I_{II} I_{III} r_1 L} - 1 \leq 0 \quad \text{(max. displ.)}, \tag{6.98}$$

$$g_2 = \frac{6 F_0 L}{X_1 X_2{}^2} - R_{p0.2} \leq 0 \quad \text{(normal stress in I)}, \tag{6.99}$$

$$g_3 = \frac{4 F_0 L}{X_3 X_4{}^2} - R_{p0.2} \leq 0 \quad \text{(normal stress in II)}, \tag{6.100}$$

$$g_4 = \frac{2 F_0 L}{X_5 X_6{}^2} - R_{p0.2} \leq 0 \quad \text{(normal stress in III)}, \tag{6.101}$$

$$g_5 = \frac{3 F_0}{2 X_1 X_2} - \frac{R_{p0.2}}{2} \leq 0 \quad \text{(shear stress in I)}, \tag{6.102}$$

$$g_6 = \frac{3 F_0}{2 X_3 X_4} - \frac{R_{p0.2}}{2} \leq 0 \quad \text{(shear stress in II)}, \tag{6.103}$$

$$g_7 = \frac{3 F_0}{2 X_5 X_6} - \frac{R_{p0.2}}{2} \leq 0 \quad \text{(shear stress in III)}, \tag{6.104}$$

$$g_8 = X_2 - 20 X_1 \leq 0 \quad \text{(height-to-width ratio in I)}, \tag{6.105}$$

$$g_9 = X_4 - 20 X_3 \leq 0 \quad \text{(height-to-width ratio in II)}, \tag{6.106}$$

$$g_{10} = X_6 - 20 X_5 \leq 0 \quad \text{(height-to-width ratio in III)}. \tag{6.107}$$

The following Listing 6.17 shows the entire wxMaxima code for the determination of the minimum of the objective function given in Eq. (6.96).

```
(% i47)   load("my_funs.mac")$

          fpprintprec:6$
          ratprint: false$
          eps : 1/1000$

          L : 2540$
          L1 : L/3$
          L2 : L/3$
          L3 : L/3$
          E : 68948$
          ro : 2.691e-6$
          R_p02 : 247$
          F_0 : 2667$
          r_1 : 0.06$

          I1 : (1/12)*X[1]*(X[2]^3)$
          I2 : (1/12)*X[3]*(X[4]^3)$
          I3 : (1/12)*X[5]*(X[6]^3)$

          u1z : 0$
          phi1y : 0$
          u2z : -4*F_0*(L^3)/(81*E*I1)$
          phi2y : 5*F_0*(L^2)/(18*E*I1)$
          u3z : -F_0*(L^3)*(23*I2+5*I1)/(162*E*I1*I2)$
          phi3y : F_0*(L^2)*(5*I2+3*I1)/(18*E*I1*I2)$
          u4z : -F_0*(L^3)*(19*I2*I3+7*I1*I3+I1*I2)/(81*E*I1*I2*I3)$
          phi4y : F_0*(L^2)*(5*I2*I3+3*I1*I3+I1*I2)/(18*E*I1*I2*I3)$
          u5z : 0$
          phi5y : 0$

          func_obj(X) := ro*L1*X[1]*X[2] + ro*L2*X[3]*X[4] + ro*L3*X[5]*X[6]$

          /* Normal Stress - Sig(x,(b/2)) */
          M_y : E*Ii*(((6/Li^2)-(12*x/Li^3))*uLz + ((-4/Li)+6*x/Li^2)*phiLy
                + ((-6/Li^2)+12*x/Li^3)*uRz + ((-2/Li)+6*x/Li^2)*phiRy)$
          Sig : M_y*(b/2)/Ii$

          /* Shear Stress - Tau(x,0) */
          Q_z : E*Ii*((-12/Li^3)*uLz + (6/Li^2)*phiLy + (12/Li^3)*uRz + (6/Li^2)*phiRy)$
          Tau : Q_z*(b/2)^2/(2*Ii)$

          /* Element 1 */
          g[1](X) := at(Sig - R_p02, [x=0, Li=L1, Ii=I1, uLz=u1z, phiLy=phi1y, uRz=u2z,
                phiRy=phi2y, b=X[2]])$
          g[2](X) := at(Tau - R_p02, [x=0, Li=L1, Ii=I1, uLz=u1z, phiLy=phi1y,
                uRz=u2z, phiRy=phi2y, b=X[2]])$
          g[3](X) := X[2] - 20*X[1]$
```

```
/* Element 2 */
g[4](X) := at(Sig - R_p02, [x=0, Li=L2, li=l2, uLz=u2z, phiLy=phi2y, uRz=u3z,
    phiRy=phi3y, b=X[4]])$
g[5](X) := at(Tau - R_p02, [x=0, Li=L2, li=l2, uLz=u2z, phiLy=phi2y,
    uRz=u3z, phiRy=phi3y, b=X[4]])$
g[6](X) := X[4] - 20*X[3]$

/* Element 3 */
g[7](X) := at(Sig - R_p02, [x=0, Li=L3, li=l3, uLz=u3z, phiLy=phi3y, uRz=u4z,
    phiRy=phi4y, b=X[6]])$
g[8](X) := at(Tau - R_p02, [x=0, Li=L3, li=l3, uLz=u3z, phiLy=phi3y,
    uRz=u4z, phiRy=phi4y, b=X[6]])$
g[9](X) := X[6] - 20*X[5]$
g[10](X) := -u4z - r_1*L$

/* Solution */
no_of_vars : 6$

X_0 : [X[1]=4,X[2]=80,X[3]=4,X[4]=80,X[5]=4,X[6]=80]$
alpha_0 : 1$
gamma_value : 1$
r_p_0s : [0.05, 0.5, 1, 10, 20, 50, 100]$

for r : 1 thru length(r_p_0s) do (
    print("==============="),
    print("For ", r_p = r_p_0s[r]),
    [X_new, pseudo_objective_fun_value] :
    Newton_multi_variable_constrained(func_obj,no_of_vars,X_0,alpha_0,eps,
        r_p_0s[r], gamma_value,"Kuhn_Tucker",true),
    print(X["1"] = rhs(X_new[1])),
    print(X["2"] = rhs(X_new[2])),
    print(X["3"] = rhs(X_new[3])),
    print(X["4"] = rhs(X_new[4])),
    print(X["5"] = rhs(X_new[5])),
    print(X["6"] = rhs(X_new[6])),
    printf(true, "The pseudo-objective function value at this point:
    F = ~,6f", pseudo_objective_fun_value),
    X_0 : copy(X_new)
)$
```

===============
For r_p=0.05
i=1 X=[X[1]=7.82797,X[2]=144.931,X[3]=7.66238,X[4]=143.592,X[5]=6.9858,X[6]=136.566] func(X)=7.26712
i=2   X=[X[1]=7.57155,X[2]=147.309,X[3]=6.78427,X[4]=127.136,X[5]=6.18523,X[6]=120.916]
func(X)=6.21804
i=3 X=[X[1]=7.50284,X[2]=147.985,X[3]=6.4824,X[4]=129.966,X[5]=5.6752,X[6]=110.945] func(X)=5.90667
i=4 X=[X[1]=7.43703,X[2]=148.641,X[3]=6.48337,X[4]=129.964,X[5]=5.2355,X[6]=102.349] func(X)=5.67982
i=5 X=[X[1]=7.43538,X[2]=148.658,X[3]=6.4834,X[4]=129.964,X[5]=5.15491,X[6]=103.137] func(X)=5.67027
i=6 X=[X[1]=7.43149,X[2]=148.697,X[3]=6.48345,X[4]=129.964,X[5]=5.15491,X[6]=103.137] func(X)=5.6697
i=7 X=[X[1]=7.43476,X[2]=148.752,X[3]=6.49496,X[4]=129.95,X[5]=5.15522,X[6]=103.144] func(X)=5.655
i=8 X=[X[1]=7.43476,X[2]=148.752,X[3]=6.49501,X[4]=129.949,X[5]=5.15522,X[6]=103.144] func(X)=5.655
Converged after 8 iterations!
X[1]=7.43476
X[2]=148.752
X[3]=6.49501
X[4]=129.949
X[5]=5.15522
X[6]=103.144
The pseudo-objective function value at this point: F = 5.654999
===============
For r_p=0.5
i=1 X=[X[1]=7.43707,X[2]=148.747,X[3]=6.4969,X[4]=129.943,X[5]=5.1566,X[6]=103.136] func(X)=5.6557
Converged after 1 iterations!
X[1]=7.43707
X[2]=148.747
X[3]=6.4969
X[4]=129.943
X[5]=5.1566
The pseudo-objective function value at this point: F = 5.655699
===============
For r_p=1
i=1 X=[X[1]=7.4372,X[2]=148.747,X[3]=6.497,X[4]=129.942,X[5]=5.15668,X[6]=103.136] func(X)=5.65574
Converged after 1 iterations!
X[1]=7.4372
X[2]=148.747
X[3]=6.497
X[4]=129.942
X[5]=5.15668
X[6]=103.136
The pseudo-objective function value at this point: F = 5.655738
===============
For r_p=10
i=1 X=[X[1]=7.43732,X[2]=148.747,X[3]=6.4971,X[4]=129.942,X[5]=5.15675,X[6]=103.135] func(X)=5.65577
Converged after 1 iterations!
X[1]=7.43732
X[2]=148.747
X[3]=6.4971
X[4]=129.942
X[5]=5.15675
X[6]=103.135
The pseudo-objective function value at this point: F = 5.655773

```
==============
For r_p=20
i=1 X=[X[1]=7.43732,X[2]=148.747,X[3]=6.4971,X[4]=129.942,X[5]=5.15675,X[6]=103.135]
func(X)=5.65577
Converged after 1 iterations!
X[1]=7.43732
X[2]=148.747
X[3]=6.4971
X[4]=129.942
X[5]=5.15675
X[6]=103.135
The pseudo-objective function value at this point: F = 5.655775
==============
For r_p=50
i=1 X=[X[1]=7.43733,X[2]=148.747,X[3]=6.4971,X[4]=129.942,X[5]=5.15675,X[6]=103.135]
func(X)=5.65578
Converged after 1 iterations!
X[1]=7.43733
X[2]=148.747
X[3]=6.4971
X[4]=129.942
X[5]=5.15675
X[6]=103.135
The pseudo-objective function value at this point: F = 5.655776
==============
For r_p=100
i=1 X=[X[1]=7.43733,X[2]=148.747,X[3]=6.4971,X[4]=129.942,X[5]=5.15676,X[6]=103.135]
func(X)=5.65578
Converged after 1 iterations!
X[1]=7.43733
X[2]=148.747
X[3]=6.4971
X[4]=129.942
X[5]=5.15676
X[6]=103.135
The pseudo-objective function value at this point: F = 5.655776
```

Module 6.17: Numerical determination of the minimum of the objective function $F(X_1, X_2, X_3, X_4, X_5, X_6)$ and the limitations due to 10 inequality constraints (see Eqs. (6.98)–(6.107))

5.8 Optimization of a Stepped Simply Supported Beam

A single stiffness matrix of an Euler-Bernoulli beam element can be stated as, see [2, 3]:

$$
\boldsymbol{K}_i^e = \frac{E_i I_i}{L_i^3}
\begin{bmatrix}
12 & -6L_i & -12 & -6L_i \\
-6L_i & 4L_i^2 & 6L_i & 2L_i^2 \\
-12 & 6L_i & 12 & 6L_i \\
-6L_i & 2L_i^2 & 6L_i & 4L_i^2
\end{bmatrix}.
\tag{6.108}
$$

Assembling the four elemental stiffness matrices $\boldsymbol{K}_i^e$ under the consideration of $E_\mathrm{I} = \cdots = E_\mathrm{IV} = E$, $L_\mathrm{I} = \cdots = L_\mathrm{IV} = \frac{L}{4}$, $I_\mathrm{IV} = I_\mathrm{I}$, and $I_\mathrm{III} = I_\mathrm{II}$ results in the following global stiffness matrix:

$$
\boldsymbol{K} =
$$

$$
E
\begin{bmatrix}
\frac{16I_1}{L} & \frac{96I_1}{L^2} & \frac{8I_1}{L} & 0 & 0 & 0 & 0 & 0 \\
\frac{96I_1}{L^2} & \left(\frac{768I_\mathrm{II}}{L^3} + \frac{768I_1}{L^3}\right) & \left(\frac{96I_1}{L^2} - \frac{96I_\mathrm{II}}{L^2}\right) & -\frac{768I_\mathrm{II}}{L^3} & \frac{96I_\mathrm{II}}{L^2} & 0 & 0 & 0 \\
\frac{8I_1}{L} & \left(\frac{96I_1}{L^2} - \frac{96I_\mathrm{II}}{L^2}\right) & \left(\frac{16I_\mathrm{II}}{L} + \frac{16I_1}{L}\right) & \frac{96I_\mathrm{II}}{L^2} & \frac{8I_\mathrm{II}}{L} & 0 & 0 & 0 \\
0 & -\frac{768I_\mathrm{II}}{L^3} & \frac{96I_\mathrm{II}}{L^2} & \frac{1536I_\mathrm{II}}{L^3} & 0 & -\frac{768I_\mathrm{II}}{L^3} & -\frac{96I_\mathrm{II}}{L^2} & 0 \\
0 & \frac{96I_\mathrm{II}}{L^2} & \frac{8I_\mathrm{II}}{L} & 0 & \frac{32I_\mathrm{II}}{L} & \frac{96I_\mathrm{II}}{L^2} & \frac{8I_\mathrm{II}}{L} & 0 \\
0 & 0 & 0 & -\frac{768I_\mathrm{II}}{L^3} & \frac{384I_\mathrm{II}}{L^3} & \left(\frac{768I_\mathrm{II}}{L^3} + \frac{768I_1}{L^3}\right) & \left(\frac{96I_\mathrm{II}}{L^2} - \frac{96I_1}{L^2}\right) & -\frac{96I_1}{L^2} \\
0 & 0 & 0 & -\frac{96I_\mathrm{II}}{L^2} & \frac{8I_\mathrm{II}}{L} & \left(\frac{96I_\mathrm{II}}{L^2} - \frac{96I_1}{L^2}\right) & \left(\frac{16I_\mathrm{II}}{L} + \frac{16I_1}{L}\right) & \frac{8I_1}{L} \\
0 & 0 & 0 & 0 & 0 & -\frac{96I_1}{L^2} & \frac{8I_1}{L} & \frac{16I_1}{L}
\end{bmatrix}.
$$

$$
\tag{6.109}
$$

The last equation already considers that the translational degrees of freedom are zero at node 1 and 5. The solution of the linear system of equations can be obtained, for example, by inverting the global stiffness matrix and multiplying with the right-hand side, i.e. $\boldsymbol{u} = \boldsymbol{K}^{-1}\boldsymbol{f}$, to obtain the column matrix of nodal unknowns:

$$
\begin{bmatrix}
\varphi_{1y} \\
u_{2z} \\
\varphi_{2y} \\
u_{3z} \\
\varphi_{3y} \\
u_{4z} \\
\varphi_{4y} \\
\varphi_{5y}
\end{bmatrix}
=
\begin{bmatrix}
\frac{F_0(I_\mathrm{II}+3I_1)L^2}{64E\,I_1 I_\mathrm{II}} \\
-\frac{F_0(2I_\mathrm{II}+9I_1)L^3}{768E\,I_1 I_\mathrm{II}} \\
\frac{3F_0L^2}{64E\,I_\mathrm{II}} \\
-\frac{F_0(I_\mathrm{II}+7I_1)L^3}{384E\,I_1 I_\mathrm{II}} \\
0 \\
-\frac{F_0(2I_\mathrm{II}+9I_1)L^3}{768E\,I_1 I_\mathrm{II}} \\
-\frac{3F_0L^2}{64E\,I_\mathrm{II}} \\
-\frac{F_0(I_\mathrm{II}+3I_1)L^2}{64E\,I_1 I_\mathrm{II}}
\end{bmatrix}.
\tag{6.110}
$$

The nodal values of the internal reactions to calculate the normal and shear stresses are summarized in Table 6.12. It should be noted here that these nodal values based on the finite element approach are in this case equal to the analytical solution.

From Table 6.12 it can be concluded that the maximum bending moment in element I and II is attained at the right-hand node and that the shear forces is constant in each element.

The normal and shear stress distributions can be generally calculated as

$$\sigma_x^e(x, z) = \frac{M_y^e(x)}{I_y} \times z, \tag{6.111}$$

$$\tau_{xz}^e(x, z) = \frac{Q_z^e(x)}{2I_y} \left[\left(\frac{b}{2} \right)^2 - z^2 \right], \tag{6.112}$$

Thus, the maximum normal stress in element I and II is attained at the right-hand node (or at the left-hand node for element III and IV) and that the shear stress is constant in each element. Thus, the critical stresses in each element can be stated as follow:

$$\sigma_{x,\mathrm{I}} = -\frac{3F_0L}{4a_\mathrm{I}\,b_\mathrm{I}^2}, \tag{6.113}$$

$$\sigma_{x,\mathrm{II}} = -\frac{3F_0L}{2a_\mathrm{II}\,b_\mathrm{II}^2}, \tag{6.114}$$

$$\sigma_{x,\mathrm{III}} = -\frac{3F_0L}{2a_\mathrm{III}\,b_\mathrm{III}^2}, \tag{6.115}$$

$$\sigma_{x,\mathrm{IV}} = -\frac{3F_0L}{4a_\mathrm{IV}\,b_\mathrm{IV}^2}, \tag{6.116}$$

or for the shear stresses:

$$\tau_{xz,\mathrm{I}} = -\frac{3F_0}{4a_\mathrm{I}\,b_\mathrm{I}}, \tag{6.117}$$

$$\tau_{xz,\mathrm{II}} = -\frac{3F_0}{4a_\mathrm{II}\,b_\mathrm{II}}, \tag{6.118}$$

$$\tau_{xz,\mathrm{III}} = \frac{3F_0}{4a_\mathrm{III}\,b_\mathrm{III}}, \tag{6.119}$$

$$\tau_{xz,\mathrm{IV}} = \frac{3F_0}{4a_\mathrm{IV}\,b_\mathrm{IV}}. \tag{6.120}$$

The objective function, i.e. the mass of the stepped beam, can be stated as a function of the four design variables $a_I = a_{IV} = X_1$, $b_I = b_{IV} = X_2$, and $a_{II} = a_{III} = X_3$, $b_{II} = b_{III} = X_4$ as:

$$F(X_1, X_2, X_3, X_2) = 2\varrho L_I X_1 X_2 + 2\varrho L_{II} X_3 X_4, \tag{6.121}$$

which is to be minimized under the following 13 inequality constraints:

$$g_1 = \frac{F_0 (I_{II} + 7I_I) L^3}{384E \, I_I I_{II} r_I L} - 1 \leq 0 \quad \text{(max. displ.)}, \tag{6.122}$$

$$g_2 = \frac{3F_0 L}{4X_1 X_2^2} - R_{p0.2} \leq 0 \quad \text{(normal stress in I)}, \tag{6.123}$$

$$g_3 = \frac{3F_0 L}{2X_3 X_4^2} - R_{p0.2} \leq 0 \quad \text{(normal stress in II)}, \tag{6.124}$$

$$g_4 = \frac{3F_0 L}{2X_3 X_4^2} - R_{p0.2} \leq 0 \quad \text{(normal stress in III)}, \tag{6.125}$$

$$g_5 = \frac{3F_0 L}{4X_1 X_2^2} - R_{p0.2} \leq 0 \quad \text{(normal stress in IV)}, \tag{6.126}$$

$$g_6 = \frac{3F_0}{4X_1 X_2} - \frac{R_{p0.2}}{2} \leq 0 \quad \text{(shear stress in I)}, \tag{6.127}$$

$$g_7 = \frac{3F_0}{4X_3 X_4} - \frac{R_{p0.2}}{2} \leq 0 \quad \text{(shear stress in II)}, \tag{6.128}$$

$$g_8 = \frac{3F_0}{4X_3 X_4} - \frac{R_{p0.2}}{2} \leq 0 \quad \text{(shear stress in III)}, \tag{6.129}$$

$$g_9 = \frac{3F_0}{4X_1 X_2} - \frac{R_{p0.2}}{2} \leq 0 \quad \text{(shear stress in IV)}, \tag{6.130}$$

$$g_{10} = X_2 - 20X_1 \leq 0 \quad \text{(height-to-width ratio in I)}, \tag{6.131}$$

$$g_{11} = X_4 - 20X_3 \leq 0 \quad \text{(height-to-width ratio in II)}, \tag{6.132}$$

$$g_{12} = X_4 - 20X_3 \leq 0 \quad \text{(height-to-width ratio in III)}, \tag{6.133}$$

$$g_{13} = X_2 - 20X_1 \leq 0 \quad \text{(height-to-width ratio in IV)}. \tag{6.134}$$

The following Listing 6.18 shows the entire wxMaxima code for the determination of the minimum of the objective function given in Eq. (6.121).

```
(% i51)   load("my_funs.mac")$

          fpprintprec:6$
          ratprint: false$
          eps : 1/1000$


          L : 2540$
          L1 : L/4$
          L2 : L/4$
          L3 : L/4$
          L4 : L/4$
          E : 68948$
          ro : 2.691e-6$
          R_p02 : 247$
          F_0 : 2667$
          r_1 : 0.01$


          I1 : (1/12)*X[1]*(X[2]^3)$
          I4 : I1$
          I2 : (1/12)*X[3]*(X[4]^3)$
          I3 : I2$


          u1z : 0$
          phi1y : F_0*(I2+3*I1)*L^2/(64*E*I1*I2)$
          u2z : -F_0*(2*I2+9*I1)*(L^3)/(768*E*I1*I2)$
          phi2y : 3*F_0*(L^2)/(64*E*I2)$
          u3z : -F_0*(I2+7*I1)*(L^3)/(384*E*I1*I2)$
          phi3y : 0$
          u4z : -F_0*(2*I2+9*I1)*(L^3)/(768*E*I1*I2)$
          phi4y : -3*F_0*(L^2)/(64*E*I2)$
          u5z : 0$
          phi5y : -F_0*(I2+3*I1)*L^2/(64*E*I1*I2)$


          func_obj(X) := 2*ro*L1*X[1]*X[2] + 2*ro*L2*X[3]*X[4]$

          /* Normal Stress - Sig(x,(b/2)) */
          M_y : E*Ii*(((6/Li^2)-(12*x/Li^3))*uLz + ((-4/Li)+6*x/Li^2)*phiLy
              + ((-6/Li^2)+12*x/Li^3)*uRz + ((-2/Li)+6*x/Li^2)*phiRy)$
          Sig : M_y*(b/2)/Ii$

          /* Shear Stress - Tau(x,0) */
          Q_z : E*Ii*((-12/Li^3)*uLz + (6/Li^2)*phiLy + (12/Li^3)*uRz + (6/Li^2)*phiRy)$
          Tau : Q_z*(b/2)^2/(2*Ii)$

          /* Element 1 */
          g[1](X) := at(Sig - R_p02, [x=L1, Li=L1, Ii=I1, uLz=u1z, phiLy=phi1y,
              uRz=u2z, phiRy=phi2y, b=X[2]])$
          g[2](X) := at(Tau - R_p02, [x=L1, Li=L1, Ii=I1, uLz=u1z, phiLy=phi1y,
              uRz=u2z, phiRy=phi2y, b=X[2]])$
          g[3](X) := X[2] - 20*X[1]$
```

```
/* Element 2 */
g[4](X) := at(Sig - R_p02, [x=L2, Li=L2, Ii=I2, uLz=u2z, phiLy=phi2y,
    uRz=u3z, phiRy=phi3y, b=X[4]])$
g[5](X) := at(Tau - R_p02, [x=L2, Li=L2, Ii=I2, uLz=u2z, phiLy=phi2y,
    uRz=u3z, phiRy=phi3y, b=X[4]])$
g[6](X) := X[4] - 20*X[3]$
g[7](X) := -u3z - r_1*L$

/* Element 3 */
g[8](X) := at(Sig - R_p02, [x=L3, Li=L3, Ii=I3, uLz=u3z, phiLy=phi3y,
    uRz=u4z, phiRy=phi4y, b=X[4]])$
g[9](X) := at(Tau - R_p02, [x=L3, Li=L3, Ii=I3, uLz=u3z, phiLy=phi3y,
    uRz=u4z, phiRy=phi4y, b=X[4]])$

/* Element 4 */
g[10](X) := at(Sig - R_p02, [x=L4, Li=L4, Ii=I4, uLz=u4z, phiLy=phi4y,
    uRz=u5z, phiRy=phi5y, b=X[2]])$
g[11](X) := at(Tau - R_p02, [x=L4, Li=L4, Ii=I4, uLz=u4z, phiLy=phi4y,
    uRz=u5z, phiRy=phi5y, b=X[2]])$

/* Solution */
no_of_vars : 4$

X_0 : [X[1]=1,X[2]=20,X[3]=1,X[4]=20]$
alpha_0 : 1$
gamma_value : 1$
r_p_0s : [0.05, 0.5, 1, 10, 20, 50, 100]$

for r : 1 thru length(r_p_0s) do (
    print("=============="),
    print("For ", r_p = r_p_0s[r]),
    [X_new, pseudo_objective_fun_value] :
    Newton_multi_variable_constrained(func_obj,no_of_vars,X_0,alpha_0,eps,
        r_p_0s[r], gamma_value,"Kuhn_Tucker",true),
    print(X["1"] = rhs(X_new[1])),
    print(X["2"] = rhs(X_new[2])),
    print(X["3"] = rhs(X_new[3])),
    print(X["4"] = rhs(X_new[4])),
    printf(true, "The pseudo-objective function value at this point:
    F = ~,6f", pseudo_objective_fun_value),
    X_0 : copy(X_new)
)$
```

```
==============
For r_p=0.05
i=1 X=[X[1]=5.24768,X[2]=104.954,X[3]=5.24768,X[4]=104.954] func(X)=3.7905
i=2 X=[X[1]=4.50338,X[2]=90.0914,X[3]=5.35487,X[4]=107.136] func(X)=3.48719
i=3 X=[X[1]=4.08859,X[2]=81.8888,X[3]=5.60818,X[4]=112.306] func(X)=3.33695
i=4 X=[X[1]=4.08226,X[2]=81.7145,X[3]=5.635,X[4]=112.796] func(X)=3.33342
i=5 X=[X[1]=4.07707,X[2]=81.6111,X[3]=5.63897,X[4]=112.876] func(X)=3.3334
Converged after 5 iterations!
X[1]=4.07707
X[2]=81.6111
X[3]=5.63897
X[4]=112.876
The pseudo-objective function value at this point: F = 3.333398
==============
For r_p=0.5
i=1 X=[X[1]=4.10215,X[2]=82.0475,X[3]=5.67365,X[4]=113.479] func(X)=3.35288
i=2 X=[X[1]=4.1021,X[2]=82.049,X[3]=5.67356,X[4]=113.481] func(X)=3.35287
Converged after 2 iterations!
X[1]=4.1021
X[2]=82.049
X[3]=5.67356
X[4]=113.481
The pseudo-objective function value at this point: F = 3.352869
==============
For r_p=1
i=1 X=[X[1]=4.10355,X[2]=82.0746,X[3]=5.67557,X[4]=113.516] func(X)=3.35399
Converged after 1 iterations!
X[1]=4.10355
X[2]=82.0746
X[3]=5.67557
X[4]=113.516
The pseudo-objective function value at this point: F = 3.353989
==============
For r_p=10
i=1 X=[X[1]=4.10487,X[2]=82.0977,X[3]=5.67739,X[4]=113.548] func(X)=3.355
i=2 X=[X[1]=4.10487,X[2]=82.0977,X[3]=5.67739,X[4]=113.548] func(X)=3.355
Converged after 2 iterations!
X[1]=4.10487
X[2]=82.0977
X[4]=113.548
The pseudo-objective function value at this point: F = 3.355000
```

```
==============
For r_p=20
i=1 X=[X[1]=4.10494,X[2]=82.099,X[3]=5.67749,X[4]=113.55] func(X)=3.35506
Converged after 1 iterations!
X[1]=4.10494
X[2]=82.099
X[3]=5.67749
X[4]=113.55
The pseudo-objective function value at this point: F = 3.355057
==============
For r_p=50
i=1 X=[X[1]=4.10499,X[2]=82.0998,X[3]=5.67755,X[4]=113.551] func(X)=3.35509
Converged after 1 iterations!
X[1]=4.10499
X[2]=82.0998
X[3]=5.67755
X[4]=113.551
The pseudo-objective function value at this point: F = 3.355091
==============
For r_p=100
i=1 X=[X[1]=4.105,X[2]=82.1,X[3]=5.67758,X[4]=113.552] func(X)=3.3551
Converged after 1 iterations!
X[1]=4.105
X[2]=82.1
X[3]=5.67758
X[4]=113.552
The pseudo-objective function value at this point: F = 3.355102
```

Module 6.18: Numerical determination of the minimum of the objective function $F(X_1, X_2, X_3, X_4)$ and the limitations due to 13 inequality constraints (see Eqs. (6.122)–(6.134))

References

1. Öchsner A (2014) Elasto-plasticity of frame structure elements: modeling and simulation of rods and beams. Springer, Berlin
2. Öchsner A (2016) Computational statics and dynamics—an introduction based on the finite element method. Springer, Singapore
3. Öchsner A (2018) A project-based introduction to computational statics. Springer, Cham
4. Öchsner A (2019) Leichtbaukonzepte anhand einfacher Strukturelemente: Neuer didaktischer Ansatz mit zahlreichen Übungsaufgaben. Springer Vieweg, Berlin

Chapter 7
Maxima Source Codes

Abstract This chapter provides the commented and structured source code of the main Maxima file which contains all the written routines.

The following file **my_funs.mac** must be included in all Maxima sheets for correct execution and result display.

A comment for cross-platform usage:

When going cross platforms (for example, from Microsoft Windows operating system to Mac OS, there is usually a problem of transferring the encoded text files to the new platform. Therefore, we recommend that users stick to UTF-8 encoding which is somehow universal. However, since older versions of the Microsoft Windows do not completely support UTF-8, it might be the case that Maxima encounters an error like:

PARSE – NAMESTRING : Character #\uF028 cannot be represented in the character set CHARSET : CP1252

In this example, the source of the problem is that the standard encoding of the operating system being used (CP-1252, also known as Windows-1252) does not support some of the characters in the UTF-8 encoded library file. In these situations, the easiest would be to convert the encoding of the library file to the encoding which is known to your destination operating system.
To easier navigate in the source code of the file **my_funs.mac**, Table 7.1 collects in alphabetical order all the functions and the corresponding line in the source code.

Table 7.1 Alphabetical list of all functions in the main library my_funs.mac

Function name	Program line
alpha_old_finder	19
alpha_old_finder_var2	73
bf_ver1	176
bf_ver2	223
bf_ver2_varN	287
bf_ver2_varN_table	360
gradient	416
gss	435
Newton_multi_variable_constrained	490
Newton_multi_variable_unconstrained	572
Newton_multi_variable_unconstrained_var2	678
Newton_one_variable_constrained_exterior_penalty	775
Newton_one_variable_constrained_exterior_penalty_var2	843
Newton_one_variable_unconstrained	897
Newton_one_variable_unconstrained_table	952
one_variable_constrained_exterior_penalty	1007
one_variable_constrained_interior_penalty	1068
one_variable_constrained_one_variable_constrained_sign_detector	1122
one_variable_constrained_range_detection	1182
pseudo_function_interior_penalty	1266
steepest_multi_variable_constrained	1313
steepest_multi_variable_unconstrained	1393
steepest_multi_variable_unconstrained_var2	1486

_____

                                                      my_funs.mac (main file)

```
1   /*
2   load("../Library/my_funs.mac");
3   */
4
5   /*
6   A routine to calculate alpha_star
7
8   Inputs:
9   func : Objective function
10  X_old_value : Solution from the previous iteration (X_0 for the first
         iteration)
11  S_old : Value of S_old
12  alpha_old_0 : alpha_star from the previous iteration (alpha_0 for the
         first iteration)
13  eps : Tolerance value
14  counter : Current iteration number
15
```

```
16  Output:
17  alpha_new : alpha_star for the current iteration
18  */
19  alpha_old_finder(func, X_old_value, S_old, alpha_old_0, eps, counter)
       :=
20  block([X_substitute, alpha_old, check],
21
22    /*assing the initial alpha to the objective function */
23    alpha_old : copy(alpha_old_0),
24
25    /* change of variables (substitution) to find X with respect to
       alpha */
26    for i:1 thru no_of_vars do X_substitute[i] :
       X[i]=X_old_value[i]+alpha*S_old[i][1],
27    X_substitute : listarray(X_substitute),
28    /* determine the objective function with respect to the unknown
       variable alpha */
29    func_alpha(X) := subst(X_substitute,func(X)),
30    dfunc : diff(func_alpha(X),alpha,1),
31    ddfunc : diff(func_alpha(X),alpha,2),
32
33    check : true,
34    j : 0,
35    while (check) do (
36      j : j + 1,
37
38      /* calculate the new alpha */
39      alpha_new : alpha_old - float(at(dfunc,
         alpha=alpha_old))/float(at(ddfunc, alpha=alpha_old)),
40
41      /* check for convergence of the solution */
42      if (abs(alpha_new-alpha_old)<eps) then (
43        check : false
44      ),
45      if (j>50) then (
46        check : false,
47        printf(true, "*** Warning, no convergence in alpha for i =
48        ~d *** ~%", counter),
49        alpha_new : alpha_old_0
50      ),
51      alpha_old : alpha_new
52    ),
53    return(alpha_new)
54  )$
55
56  /*
57  A routine to calculate alpha_star
58  Variation 2
59
60  Purpose:
61  This function works the same as *alpha_old_finder*, with the
       difference that the pseudo-objective function with respect to
       alpha, and its gradients must be calculated separately within the
       routine.
```

```
62
63  Inputs:
64  func_obj : Objective function
65  X_old_value : Solution from the previous iteration (X_0 for the first
        iteration)
66  S_old : Value of S_old
67  alpha_old_0 : alpha_star from the previous iteration (alpha_0 for the
        first iteration)
68  eps : Tolerance value
69  counter : Current iteration number
70
71  Output:
72  alpha_new : alpha_star for the current iteration
73  */
74  alpha_old_finder_var2(func_obj, X_old_value, S_old, alpha_old_0, eps,
        counter) :=
75  block([X_substitute, alpha_old, check, linear_of, p_alpha, dp_alpha,
        ddp_alpha, g_alpha],
76
77    /*assing the initial alpha to the function */
78    alpha_old : copy(alpha_old_0),
79
80    /* change of variables (substitution) to find X with respect to
        alpha */
81    for i:1 thru no_of_vars do X_substitute[i] :
        X[i]=X_old_value[i]+alpha*S_old[i][1],
82    X_substitute : listarray(X_substitute),
83
84    /* determine the objective function with respect to the unknown
        variable alpha */
85    func_alpha_subs : subst(X_substitute,func_obj(X)),
85
86
87      /* derive the function and its derivatives with respect to alpha */
88      /* if g and h exist, calculate p(x) and q(x) */
89      if (unknown(g[1])) then (
90       len_g : length(listarray(g)),
91       for n:1 thru len_g do (
92         g_alpha[n] : subst(X_substitute,g[n](X))
93       ),
94       p_alpha : sum(unit_step(g_alpha[m])*g_alpha[m]^2,m,1,len_g),
95       dp_alpha :
      sum(unit_step(g_alpha[m])*diff(g_alpha[m]^2,alpha,1),m,1,len_g),
96       ddp_alpha :
      sum(unit_step(g_alpha[m])*diff(g_alpha[m]^2,alpha,2),m,1,len_g)
97      ),
98
99      if (unknown(h[1])) then (
100      len_h : length(listarray(h)),
101      for k:1 thru len_h do (
102        h_alpha[k] : subst(X_substitute,h[k](X))
103      ),
104      q_alpha : sum(h_alpha[k]^2,k,1,len_h),
105      dq_alpha : sum(diff(h_alpha[k]^2,alpha,1),k,1,len_h),
```

```
106        ddq_alpha : sum(diff(h_alpha[k]^2,alpha,2),k,1,len_h)
107      ),
108
109      func_alpha : func_alpha_subs,
110      dfunc_alpha : diff(func_alpha_subs,alpha,1),
111      ddfunc_alpha : diff(func_alpha_subs,alpha,2),
112      r_p : r_p_0,
113      if ( unknown(g[1]) and unknown(h[1]) ) then (
114        func_alpha : func_alpha_subs + r_p*p_alpha + r_p*q_alpha,
115        dfunc_alpha : diff(func_alpha_subs,alpha,1) + r_p*dp_alpha +
         r_p*dq_alpha,
116        ddfunc_alpha : diff(func_alpha_subs,alpha,2) + r_p*ddp_alpha +
         r_p*ddq_alpha
117      ) else if ( unknown(g[1]) and not(unknown(h[1])) ) then (
118        func_alpha : func_alpha_subs + r_p*p_alpha,
119        dfunc_alpha : diff(func_alpha_subs,alpha,1) + r_p*dp_alpha,
120        ddfunc_alpha : diff(func_alpha_subs,alpha,2) + r_p*ddp_alpha
121      ) else if ( not(unknown(g[1])) and unknown(h[1]) ) then (
122        func_alpha : func_alpha_subs + r_p*q_alpha,
123        dfunc_alpha : diff(func_alpha_subs,alpha,1) + r_p*dq_alpha,
124        ddfunc_alpha : diff(func_alpha_subs,alpha,2) + r_p*ddq_alpha
125      ),
126
127    check : true,
128    j : 0,
129    while (check) do (
130      linear_of : false,
131      if (at(float(ddfunc_alpha), alpha=alpha_old)=0) then (
132        linear_of : true
133      ),
134      j : j + 1,
135
136      /* calculate the new alpha */
137      if not(linear_of) then (
138        alpha_new : alpha_old - float(at(dfunc_alpha,
         alpha=alpha_old))/float(at(float(ddfunc_alpha), alpha=alpha_old))
139      ) else (
140        alpha_new : copy(alpha_old/1000),
141        check : false
142        /*
143        if (at(func_alpha,alpha=1E-3)<=at(func_alpha,alpha=1-1E-3))
         then (
144          alpha_new : 1E-3
145        ) else (
146          alpha_new : 1-1E-3
147        )
148        */
149      ),
150
151      /* check for convergence of the solution */
152      if (abs(float(at(dfunc_alpha, alpha=alpha_new)))<eps) then (
153        check : false
154      ),
155      if (j>50 and check) then (
```

```
156        check : false,
157        printf(true, "*** Warning, no convergence in alpha for i = ~d
     *** ~%", counter),
158          alpha_new : copy(alpha_old_0)
159        ),
160      alpha_old : copy(alpha_new)
161    ),
162    return(alpha_new)
163  )$
164
165  /*
166  Brute Force Method (version 1)
167
168  Inputs:
169  Xmin : Minimum value of X
170  Xmax : Maximum value of X
171  N : Total number of function evaluation
172
173  Output:
174
175  */
176  bf_ver1(Xmin, Xmax, N) :=
177  block([X0, X1, X2, X_extr, fX0, FX1, FX2, Fmin, Fmax, dh, i],
178    Fmin : func(Xmin),
179    Fmax : func(Xmax),
180    X0 : Xmin,
181    dh : (Xmax - Xmin) / N,
182    i : 0,
183    while true do (
184      i : i + 1,
185      X1 : X0 + dh,
186      X2 : X1 + dh,
187      fX0 : func(X0),
188      FX1 : func(X1),
189      FX2 : func(X2),
190      if ( (fX0>=FX1) and (FX1<=FX2) ) then (
191        X_extr : (X0 + X2) / 2,
192        printf(true, "~% minimum lies in [~12,4e,~12,4e]", X0, X2),
193        printf(true, "~% X_extr = ~12,4e ( i = ~d )", X_extr, i),
194        return(X_extr)
195      )
196      else (
197        if ( (X2 < Xmax) ) then (
198          X0 : X1
199        )
200        else (
201          printf(true, "~% no minimum lies in [~12,4e,~12,4e]", Xmin,
     Xmax),
202          printf(true,
203          "~% or boundary point (Xmin = ~4,4e ( f(Xmin)=~4,4e ) or Xmax
     = ~4,4e ( f(Xmax)=~4,4e )) is the minimum.",Xmin, Fmin, Xmax,
     Fmax),
204          return()
205        )
```

```
206        )
207      ),
208      return()
209    )$
210
211    /*
212    Brute Force Method (version 2)
213
214    Inputs:
215    Xmin : Minimum value of X
216    Xmax : Maximum value of X
217    X0 : Start value
218    N : Total number of function evaluation
219
220    Output:
221
222    */
223    bf_ver2(Xmin, Xmax, X0, n) :=
224    block([X1, X2, X_extr, fX0, FX1, FX2, Fmin, Fmax, dh, nnew, i],
225      nnew : n,
226      check : true,
227      while (check) do (
228        dh : (Xmax - Xmin) / nnew,
229        X1 : X0 - dh,
230        X2 : X0 + dh,
231        if ( (X1>=Xmin) and (X2<=Xmax) ) then (
232          check : false,
233          i : 0,
234          while true do (
235            i : i + 1,
236            X1 : X0 - dh,
237            X2 : X0 + dh,
238            fX0 : func(X0),
239            FX1 : func(X1),
240            FX2 : func(X2),
241            if ( (FX1>=fX0) and (fX0<=FX2) ) then (
242              X_extr : (X0 + X2) / 2,
243              printf(true, "~% minimum lies in [~12,4e,~12,4e]", X0, X2),
244              printf(true, "~% X_extr = ~12,4e ( i = ~d )", X_extr, i),
245              return(X_extr)
246            ) else (
247              if (FX2<=fX0) then (
248                if not(X2<Xmax) then (
249                  X_extr : Xmax,
250                  printf(true, "~% X_extr = ~12,4e ( i = ~d )", X_extr, i),
251                  return(X_extr)
252                ),
253                X0 : X2
254              ) else (
255                if not(X1>Xmin) then (
256                  X_extr : Xmin,
257                  printf(true, "~% X_extr = ~12,4e ( i = ~d )", X_extr, i),
258                  return(X_extr)
259                ),
```

```
260              X0 : X1
261            )
262          )
263        )
264      ) else (
265        nnew : nnew * 10,
266        printf(true, "~% the value of n has been changed to ~4,3f", nnew)
267      )
268    ),
269    return()
270  )$
271
272  /*
273  Brute Force Method (version 2 - with variable N)
274
275  Inputs:
276  Xmin : Minimum value of X
277  Xmax : Maximum value of X
278  X0 : Start value
279  N : Total number of function evaluation
280  alpha : Scaling parameter
281  print_swtich : Prints the output if TRUE
282
283  Output:
284  X_extr : The extremum value
285  k : Total number of iterations
286  */
287  bf_ver2_varN(Xmin, Xmax, X0, n, alpha, print_switch) :=
288  block([X1, X2, fX0, FX1, FX2, dh],
289    check : true,
290    dh : (Xmax - Xmin) / n,
291    k : 0,
292    while (true) do (
293      X1 : X0 - dh,
294      X2 : X0 + dh,
295      if ( (X1>=Xmin) and (X2<=Xmax) ) then (
296        k : k + 1,
297        fX0 : func(X0),
298        FX1 : func(X1),
299        FX2 : func(X2),
300        if ( (FX1>=fX0) and (fX0<=FX2) ) then (
301          X_extr : (X0 + X2) / 2,
302          if (print_switch) then (
303            printf(true, "~%~5d ~12,8e ~5d", n, X_extr, k)
304          ),
305          return([X_extr,k])
306        ) else (
307          if (FX2<=fX0) then (
308            if not(X2<Xmax) then (
309              X_extr : Xmax,
310              if (print_switch) then (
311                printf(true, "~5d ~12,8e ~5d", n, X_extr, k)
312              ),
313              return([X_extr,k])
```

```
314                    ),
315                    X0 : X2
316                  ) else (
317                    if not(X1>Xmin) then (
318                      X_extr : Xmin,
319                      if (print_switch) then (
320                        printf(true, "~%5d ~12,8e ~5d", n, X_extr, k)
321                      ),
322                      return([X_extr,k])
323                    ),
324                    X0 : X1
325                  )
326                ),
327                if (is(equal(alpha,"Fibonacci"))) then (
328                  dh : dh * fib(k)
329                ) else (
330                  dh : dh * alpha
331                ),
332                if (print_switch) then (
333                  printf(true, "~% New dh: ~12,8e ", dh)
334                )
335            ) else (
336              dh : dh / 100,
337              if (print_switch) then (
338                printf(true, "~% dh too large, the value has decreased to
           ~12,8e", dh)
339              )
340            )
341          ),
342        return([X_extr,k])
343      )$
344
345   /*
346   Brute Force Method (version 2 - with variable N)
347   with table output (see Module 2.6)
348
349   Inputs:
350   Xmin : Minimum value of X
351   Xmax : Maximum value of X
352   X0 : Start value
353   N : Total number of function evaluation
354   alpha : Scaling parameter
355
356   Output:
357   X_extr : The extremum value
358   k : Total number of iterations
359   */
360   bf_ver2_varN_table(Xmin, Xmax, X0, n, alpha) :=
361   block([X1, X2, X_extr, fX0, FX1, FX2, dh, nnew],
362      nnew : n,
363      dh : (Xmax - Xmin) / nnew,
364      k : 0,
365      while (true) do (
366        X1 : X0 - dh,
```

```
367        X2 : X0 + dh,
368        if ( (X1>=Xmin) and (X2<=Xmax) ) then (
369            k : k + 1,
370            fX0 : func(X0),
371            FX1 : func(X1),
372            FX2 : func(X2),
373            if ( (FX1>=fX0) and (fX0<=FX2) ) then (
374              X_extr : (X0 + X2) / 2,
375              printf(true, "~%~5d ~12,8e ~5d", n, X_extr, k),
376              return(X_extr)
377            ) else (
378              if (FX2<=fX0) then (
379                if not(X2<Xmax) then (
380                  X_extr : Xmax,
381                  printf(true, "~5d ~12,8e ~5d", n, X_extr, k),
382                  return(X_extr)
383                ),
384                X0 : X2
385              ) else (
386                if not(X1>Xmin) then (
387                  X_extr : Xmin,
388                  printf(true, "~%5d ~12,8e ~5d", n, X_extr, k),
389                  return(X_extr)
390                ),
391                X0 : X1
392              )
393            ),
394            if (equal(alpha,"Fibonacci")) then (
395              dh : dh * fib(k)
396            ) else (
397              dh : dh * alpha
398            )
399        ) else (
400            dh : dh / 100
401        )
402      ),
403      return()
404  )$
405
406  /*
407  Gradient operator
408
409  Inputs:
410  f  : Function based on XI
411  XI : Vector of variables
412
413  Output:
414  grad : Gradient of the function
415  */
416  gradient(f, XI) :=
417  block([],
418      for i : 1 thru length(XI) do (
419        grad[i] : diff(f,XI[i],1)
420      ),
```

```
421    return(listarray(grad))
422  )$
423
424  /*
425  Golden Section Search Method
426
427  Inputs:
428  Xmin : Minimum value of X
429  Xmax : Maximum value of X
430  N : Total number of function evaluation
431
432  Output:
433
434  */
435  gss(Xmin, Xmax, N):=
436  block([X1, X2, FX1, FX2, Fmin, Fmax, tau, K],
437    tau : 0.381966,
438    Fmin : func(Xmin),
439    Fmax : func(Xmax),
440    X1 : Xmin + tau*(Xmax - Xmin),
441    X2 : Xmax - tau*(Xmax - Xmin),
442    FX1 : func(X1),
443    FX2 : func(X2),
444    K : 3,
445    printf(true,"~<~%~7a~>","K"),
446    printf(true,"~{~12a~}",
447       ["X_min","X_1","X_2","X_max","f_min","f_1","f_2","f_max"]),
447    while true do (
448      printf(true,"~%~4d",K),
449      printf(true,"~{~12,4e~}",
450      [float(Xmin),float(X1),float(X2),float(Xmax),float(Fmin),
450      float(FX1),float(FX2),float(Fmax)]),
451      K : K + 1,
452      if (K > N) then return(),
453      if (FX1 > FX2) then (
454        Xmin : X1,
455        Fmin : FX1,
456        X1 : X2,
457        FX1 : FX2,
458        X2 : Xmax - tau*(Xmax - Xmin),
459        FX2 : func(X2)
460      ) else (
461        Xmax : X2,
462        Fmax : FX2,
463        X2 : X1,
464        FX2 : FX1,
465        X1 : Xmin + tau*(Xmax - Xmin),
466        FX1 : func(X1)
467      )
468    ),
469    return()
470  )$
471
472  /*
```

```
473  Constrained functions of several variables
474  Using Newton's Method
475  The exterior penalty function method
476
477  Inputs:
478  func_obj : Objective function
479  no_of_vars : Total number of variables
480  X_0 : Vector of start values
481  alpha_0 : Initial scaling parameter (alpha)
482  gamma_value : Value of the scaling parameter (gamma)
483  eps : Tolerance value
484  criterion : Convergence criterion  (max_iter, abs_change, rel_change
         or Kuhn_Tucker)
485  print_switch : Prints the output if true
486
487  Output:
488  x_extr_new : The extremum value
489  f_extr_new : The function value at the extremum
490  */
491  Newton_multi_variable_constrained(func_obj,no_of_vars,X_0,alpha_0,
492  eps,r_p_0, gamma_value,criterion,print_switch) :=
493  block([i, r_p, X_new_value, X_old_value, p, p_grad, p_hess, g, g_grad,
         g_hess],
494
495     for i:1 thru no_of_vars do X[i] : X[i],
496
497     /* if g and h exist, calculate p(x) and q(x) */
498     if (unknown(g[1])) then (
499        len_g : length(listarray(g)),
500        p : sum(unit_step(g[m](X))*g[m](X)^2,m,1,len_g),
501        p_grad : sum(unit_step(g[m](X))*gradient(g[m](X)^2,
         listarray(X)),m,1,len_g),
502        p_hess : sum(unit_step(g[m](X))*hessian(g[m](X)^2,
         listarray(X)),m,1,len_g)
503     ),
504
505     if (unknown(h[1])) then (
506        len_h : length(listarray(h)),
507        q : sum(h[k](X)^2,k,1,len_h),
508        q_grad : sum(gradient(h[k](X)^2,listarray(X)),k,1,len_h),
509        q_hess : sum(hessian(h[k](X)^2,listarray(X)),k,1,len_h)
510     ),
511
512     /* determine the pseudo-objective function */
513     func : func_obj(X),
514     func_grad : gradient(func, listarray(X)),
515     func_hess : hessian(func, listarray(X)),
516
517     check_constrained : true, /* convergence check variable for the
         constrained problem */
518     x_extr_old : copy(X_0),
519     r_p : r_p_0,
520     while (check_constrained) do (
521        if ( unknown(g[1]) and unknown(h[1]) ) then (
```

```
522       func : func_obj(X) + r_p*p + r_p*q,
523       func_grad : gradient(func_obj(X), listarray(X)) + r_p*p_grad +
          r_p*q_grad,
524       func_hess : hessian(func_obj(X), listarray(X)) + r_p*p_hess +
          r_p*q_hess
525     ) else if ( unknown(g[1]) and not(unknown(h[1])) ) then (
526       func : func_obj(X) + r_p*p,
527       func_grad : gradient(func_obj(X), listarray(X)) + r_p*p_grad,
528       func_hess : hessian(func_obj(X), listarray(X)) + r_p*p_hess
529     ) else if ( not(unknown(g[1])) and unknown(h[1]) ) then (
530       func : func_obj(X) + r_p*q,
531       func_grad : gradient(func_obj(X), listarray(X)) + r_p*q_grad,
532       func_hess : hessian(func_obj(X), listarray(X)) + r_p*q_hess
533     ),
534
535     /* send the problem as an unconstrained one to the corresponding
          solving routine */
536     [counter,x_extr_new,f_extr_new] :
          Newton_multi_variable_unconstrained_var2(func_obj,func,func_grad,
537       func_hess,no_of_vars,X_0,alpha_0,eps,criterion,print_switch),
538
539     /* check for convergence of the solution */
540     kill(X_old_value,X_new_value),
541     for i:1 thru no_of_vars do X_new_value[i] : X[i] =
          rhs(x_extr_new[i]),
542     for i:1 thru no_of_vars do X_old_value[i] : X[i] =
          rhs(x_extr_old[i]),
543     X_new_value : listarray(X_new_value),
544     X_old_value : listarray(X_old_value),
545     f_new_constrained : at(func,X_new_value),
546     f_old_constrained : at(func,X_old_value),
547     if ( abs(f_new_constrained - f_old_constrained) < eps or
          is(gamma_value=1) ) then (
548       check_constrained : false
549     ),
550
551     x_extr_old : copy(x_extr_new),
552
553     r_p : r_p * gamma_value
554   ),
555   for i:1 thru no_of_vars do x_extr_new[i] : X[i] = rhs(x_extr_new[i]),
556
557   return([x_extr_new,f_extr_new])
558 )$
559
560 /*
561 Unconstrained functions of several variables
562 Using Newton's Method
563
564 Inputs:
565 func : Objective function
566 no_of_vars : Total number of variables
567 X_0 : Vector of start values
568 eps : Tolerance value
```

```
569   criterion : Convergence criterion   (max_iter, abs_change, rel_change
          or Kuhn_Tucker)
570   print_switch : Prints the output if true
571
572   Output:
573   X_new : The extremum value
574   */
575   Newton_multi_variable_unconstrained(func,no_of_vars,X_0,alpha_0,eps,
576   criterion,print_switch) :=
577   block([i, X_new, X_new_value, X_old_value, counter],
578
579     for i:1 thru no_of_vars do X[i] : X[i],
580
581     /* calculate the gradient of the function (included in this
          libraray) */
582     f_grad : gradient(func(X), listarray(X)),
583
584     /* calculate the hessian of the function (uses a built-in maxima
          function)*/
585     f_hessian : hessian(func(X), listarray(X)),
586     X_old : copy(X_0),
587
588     check : true,
589     once : true,
590     counter : 0,
591     while (check) do (
592       counter : counter + 1,
593
594     /* clean memory from old values */
595       kill(X_old_value,X_new_value,X_new),
596
597     /* assign X_old_value */
598       for i:1 thru no_of_vars do X_old_value[i] : rhs(X_old[i]),
599       X_old_value : transpose(listarray(X_old_value)),
600
601     /* calculate the gradient, the Hessian, and S_old*/
602       f_hessian_value : float(at(f_hessian, X_old)),
603     /* make sure that the hessian has no "diagonal" zero values */
604       if (once) then (
605         for j : 1 thru no_of_vars do (
606           if (f_hessian_value[j,j]=0.0) then (
607             f_hessian_value[j,j] : 1E-6
608           )
609         ),
610       once : false
611       ),
612       f_grad_value : float(at(f_grad, X_old)),
613       /* calculate the invert of hessian */
614       f_hessian_inv_value : invert(f_hessian_value),
615       S_old : (-1*f_hessian_inv_value).f_grad_value,
616
617     /* find alpha_star */
618       alpha_star : alpha_old_finder(func, X_old_value, S_old, alpha_0,
          eps, counter),
```

```
619
620     /* calculate X_new_value */
621       for i:1 thru no_of_vars do X_new[i] : X_old[i] +
        alpha_star*S_old[i][1],
622       X_new : listarray(X_new),
623       for i:1 thru no_of_vars do X_new_value[i] : X[i] = rhs(X_new[i]),
624       X_new_value : listarray(X_new_value),
625
626     /* check if the solution is converged */
627       f_old : at(func(X),X_old),
628       f_new : at(func(X),X_new_value),
629       if (print_switch) then (
630         print(i=counter,X=X_new_value,"func(X)"=f_new)
631       ),
632       /* convergence criterion */
633       if (criterion = "max_iter") then (
634         if (counter=maximum_iteration) then check : false
635       ) elseif (criterion = "abs_change") then (
636         if (abs(f_new-f_old)<eps) then check : false
637       ) elseif (criterion = "rel_change") then (
638         if ((abs(f_new-f_old)/max(f_new,10^-10))<eps) then check : false
639       ) elseif (criterion = "Kuhn_Tucker") then (
640         f_grad_value_new : float(at(f_grad, X_new_value)),
641         check : false,
642         for i:1 thru no_of_vars do (
643           if (abs(f_grad_value_new[i])>eps) then check : true
644         )
645       ),
646
647       /* update X_old if the solution is not converged*/
648       if (check) then (
649         for i:1 thru no_of_vars do X_old[i] : X[i] = rhs(X_new[i]),
650         alpha_0 : copy(alpha_star)
651       )
652     ),
653     printf(true,"Converged after ~d iterations!", counter),
654     return(X_new)
655   )$
656
657 /*
658 Unconstrained functions of several variables
659 Using Newton's Method
660 Variation 2
661
662 Purpose:
663 This function works the same as *Newton_multi_variable_unconstrained*,
      with the difference that now the pseudo-objective function is
      also an input
664
665 Inputs:
666 func_obj : Objective function
667 func : Pseudo-objective function
668 func_grad : Gradient of the pseudo-objective function
669 func_hess : Hessian of the pseudo-objective function
```

```
670   no_of_vars : Total number of variables
671   X_0 : Vector of start values
672   alpha_0 : Initial scaling parameter (alpha)
673   eps : Tolerance value
674   criterion : Convergence criterion  (max_iter, abs_change, rel_change
         or Kuhn_Tucker)
675   print_switch : Prints the output if true
676
677   Output:
678   counter : Total number of iterations required for convergence
679   X_new : The extremum value
680   f_new : The function value at the extremum
681   */
682   Newton_multi_variable_unconstrained_var2(func_obj,func,func_grad,
683   func_hess,no_of_vars,X_0,alpha_0,eps,criterion,print_switch) :=
684   block([i, X_new, X, alpha_star],
685
686     for i:1 thru no_of_vars do X[i] : X[i],
687
688     X_old : copy(X_0),
689
690     alpha_0_store : copy(alpha_0),
691
692     check : true,
693     once : true,
694     counter : 0,
695
696     while (check) do (
697       counter : counter + 1,
698
699     /* clean memory from old values */
700       kill(X_old_value,X_new_value,X_new,S_old),
701
702     /* assign X_old_value */
703       for i:1 thru no_of_vars do X_old_value[i] : rhs(X_old[i]),
704       X_old_value : transpose(listarray(X_old_value)),
705
706     /* calculate the gradient, the Hessian and S_old*/
707       f_hessian_value : float(at(func_hess, X_old)),
708       if (once) then (
709         for j : 1 thru no_of_vars do (
710           if (f_hessian_value[j,j]=0.0) then (
711             f_hessian_value[j,j] : 1E-6
712           )
713         ),
714       once : false
715       ),
716       f_grad_value : float(at(func_grad, X_old)),
717       /* calculate the invert of hessian */
718       f_hessian_inv_value : invert(f_hessian_value),
719       S_old : (-1*f_hessian_inv_value).f_grad_value,
720
721     /* find alpha_star */
```

```
722    alpha_star : alpha_old_finder_var2(func_obj, X_old_value, S_old,
       alpha_0_store, eps, counter),
723
724  /* calculate X_new_value */
725    for i:1 thru no_of_vars do X_new[i] : X_old[i] +
       alpha_star*S_old[i][1],
726    X_new : listarray(X_new),
727    for i:1 thru no_of_vars do X_new_value[i] : X[i] = rhs(X_new[i]),
728    X_new_value : listarray(X_new_value),
729
730  /* check if the solution is converged */
731    f_old : at(func,X_old),
732    f_new : at(func,X_new_value),
733    if (print_switch) then (
734      print(i=counter,X=X_new_value,"func(X)"=float(f_new))
735    ),
736    /* convergence criterion */
737    if (criterion = "max_iter") then (
738      if (counter=maximum_iteration) then check : false
739    ) elseif (criterion = "abs_change") then (
740      if (abs(f_new-f_old)<eps) then check : false
741    ) elseif (criterion = "rel_change") then (
742      if ((abs(f_new-f_old)/max(f_new,10^-10))<eps) then check : false
743    ) elseif (criterion = "Kuhn_Tucker") then (
744      f_grad_value_new : float(at(func_grad, X_new_value)),
745      check : false,
746      for i:1 thru no_of_vars do (
747        if (abs(f_grad_value_new[i])>eps) then
748        (
749        check : true
750        )
751      )
752    ),
753
754    /* update X_old if the solution is not converged*/
755    if (check) then (
756      for i:1 thru no_of_vars do X_old[i] : X[i] = rhs(X_new[i]),
757      alpha_0_store : copy(alpha_star)
758    )
759  ),
760  printf(true,"Converged after ~d iterations!", counter),
761  return([counter,X_new,f_new])
762  )$
763
764  /*
765  Newton's Method (variation 1)
766  Constrained functions of one variable
767  The exterior penalty function method
768
769  Inputs:
770  Xmin : Minimum value of X
771  Xmax : Maximum value of X
772  X0 : Start value
773  eps : Tolerance
```

```
774   r_p_0 : Initial penalty factor
775   gamma_value : Scaling parameter for r_p
776
777   Output:
778   X_extr_new : The extremum value
779   */
780   Newton_one_variable_constrained_exterior_penalty(Xmin, Xmax, X0, eps,
          r_p_0, gamma_value) :=
781   block([r_p],
782     /* tolerance */
783     tol : 0.001,
784
785     /* if g and h exist, calculate p(x) and q(x) */
786     if (unknown(g[1])) then (
787       len_g : length(listarray(g)),
788       p(x) := sum(unit_step(g[m](x))*g[m](x)^2,m,1,len_g),
789       dp(x) := sum(unit_step(g[m](x)*diff(g[m](x)^2,x,1),m,1,len_g),
790       ddp(x) := sum(unit_step(g[m](x))*diff(g[m](x)^2,x,2),m,1,len_g)
791     ),
792     if (unknown(h[1])) then (
793       len_h : length(listarray(h)),
794       q(x) := sum(h[k](x)^2,k,1,len_g),
795       dq(x) := sum(diff(h[k](x)^2,x,1),k,1,len_g),
796       ddq(x) := sum(diff(h[k](x)^2,x,2),k,1,len_g)
797     ),
798     func(x) := f(x),
799     dfunc(x) := diff(f(x),x,1),
800     ddfunc(x) := diff(f(x),x,2),
801     check : true,
802     X_extr_old : X0,
803     r_p : r_p_0,
804     while (check) do (
805       if ( unknown(g[1]) and unknown(h[1]) ) then (
806         kill(func,dfunc,ddfunc),
807         func(x) := f(x) + r_p*p(x) + r_p*q(x),
808         dfunc(x) := diff(f(x),x,1) + r_p*dp(x) + r_p*dq(x),
809         ddfunc(x) := diff(f(x),x,2) + r_p*ddp(x) + r_p*ddq(x)
810       ) else if ( unknown(g[1]) and not(unknown(h[1])) ) then (
811         kill(func,dfunc,ddfunc),
812         func(x) := f(x) + r_p*p(x),
813         dfunc(x) := diff(f(x),x,1) + r_p*dp(x),
814         ddfunc(x) := diff(f(x),x,2) + r_p*ddp(x)
815       ) else if ( not(unknown(g[1])) and unknown(h[1]) ) then (
816         kill(func,dfunc,ddfunc),
817         func(x) := f(x) + r_p*q(x),
818         dfunc(x) := diff(f(x),x,1) + r_p*dq(x),
819         ddfunc(x) := diff(f(x),x,2) + r_p*ddq(x)
820       ),
821       [X_extr_new,n_iter] :
          Newton_one_variable_constrained_exterior_penalty_var2(Xmin, Xmax,
          X0, eps, false),
822       printf(true, "~% r_p: ~,6f , X_extr: ~,6f , Number of iterations:
          ~d", r_p, float(X_extr_new), n_iter),
```

```
823        if ( abs(X_extr_new - X_extr_old) < tol or is(gamma_value=1) )
           then (
824            check : false
825          ),
826          X_extr_old : X_extr_new,
827          r_p : r_p * gamma_value
828        ),
829      return(X_extr_new)
830  )$
831
832  /*
833  Newton's Method (variation 2)
834  Constrained functions of one variable
835  The exterior penalty function method
836
837  Inputs:
838  Xmin : Minimum value of X
839  Xmax : Maximum value of X
840  X0 : Start value
841  eps : Tolerance
842  print_swtich : Prints the output if TRUE
843
844  Output:
845  X_extr : The extremum value
846  i : Total number of iterations
847  */
848  Newton_one_variable_constrained_exterior_penalty_var2(Xmin, Xmax, X0,
         eps, print_switch) :=
849  block([Xold, Xnew, i, linear_of, X_extr],
850    linear_of : false,
851    if (ddfunc(X)=0) then ( linear_of : true ),
852    if (linear_of) then (
853      i : 1,
854      if (func(Xmin)<=func(Xmax)) then (
855        X_extr : Xmin
856      ) else (
857        X_extr : Xmax
858      )
859    ) else (
860    i : 0,
861    Xold : X0,
862    check : true,
863    while (check) do (
864      i : i + 1,
865      Xnew : Xold -
         bfloat(at(ddfunc(X),X=Xold))^(-1)*bfloat(at(dfunc(X),X=Xold)),
866      if (abs(Xnew-Xold)<eps) then (
867        X_extr : Xnew,
868        check : false
869      ) else (
870        if (Xnew <= Xmax and Xnew>=Xmin) then (
871          Xold : Xnew
872        ) else (
873          if (func(Xmin)<=func(Xmax)) then (
```

```
874          X_extr : Xmin
875        ) else (
876          X_extr : Xmax
877        ),
878        check : false
879      )
880    )
881  ) ),
882  if (print_switch) then (
883    printf(true, "~% X_extr = ~12,4e ( i = ~d )", float(X_extr), i)
884  ),
885  return([X_extr, i])
886 )$
887
888 /*
889 Newton's Method for an unconstrained minimum
890
891 Inputs:
892 Xmin : Minimum value of X
893 Xmax : Maximum value of X
894 X0 : Start value
895 eps : Tolerance
896 print_swtich : Prints the output if TRUE
897
898 Output:
899 X_extr : The extremum value
900 i : Total number of iterations
901 */
902 Newton_one_variable_unconstrained(Xmin, Xmax, X0, eps, print_switch) :=
903 block([Xold, Xnew, i, linear_of],
904   linear_of : false,
905   if (diff(func(X),X,2)=0) then ( linear_of : true ),
906   if (linear_of) then (
907     i : 1,
908     if (func(Xmin)<=func(Xmax)) then (
909       X_extr : Xmin
910     ) else (
911       X_extr : Xmax
912     )
913   ) else (
914   i : 0,
915   Xold : X0,
916   check : true,
917   while (check) do (
918     i : i + 1,
919     Xnew : Xold - bfloat(at(diff(func(X),X,2),X=Xold))^(-1)*bfloat
920     (at(diff(func(X),X,1),X=Xold)),
921     if (abs(Xnew-Xold)<eps) then (
922       X_extr : Xnew,
923       check : false
924     ) else (
925       if (Xnew <= Xmax and Xnew>=Xmin) then (
926         Xold : Xnew
927       ) else (
```

```
928          if (func(Xmin)<=func(Xmax)) then (
929            X_extr : Xmin
930          ) else (
931            X_extr : Xmax
932          ),
933          check : false
934        )
935      )
936    ) ),
937    if (print_switch) then (
938      printf(true, "~% X_extr = ~12,4e ( i = ~d )", float(X_extr), i)
939    ),
940    return([X_extr, i])
941  )$
942
943  /*
944  Newton's Method for an unconstrained minimum
945  with table output (see Module 2.11)
946
947  Inputs:
948  Xmin : Minimum value of X
949  Xmax : Maximum value of X
950  X0  : Start value
951  eps : Tolerance
952  print_swtich : Prints the output if TRUE
953
954  Output:
955  X_extr : The extremum value
956  i : Total number of iterations
957  */
958  Newton_one_variable_unconstrained_table(Xmin, Xmax, X0, eps) :=
959  block([Xold, Xnew, i, linear_of],
960    linear_of : false,
961    if (diff(func(X),X,2)=0) then ( linear_of : true ),
962    if (linear_of) then (
963      i : 1,
964      if (func(Xmin)<=func(Xmax)) then (
965        X_extr : Xmin
966      ) else (
967        X_extr : Xmax
968      )
969    ) else (
970    i : 0,
971    Xold : X0,
972    printf(true, "~%~5d  ~,8f  ~,8f", i, Xold, at(func(X),X=Xold)),
973    check : true,
974    while (check) do (
975      i : i + 1,
976      Xnew : Xold -
      at(diff(func(X),X,2),X=Xold)^(-1)*at(diff(func(X),X,1),X=Xold),
977      printf(true, "~%~5d  ~,8f  ~,8f", i, Xnew, at(func(X),X=Xnew)),
978      if (abs(Xnew-Xold)<eps) then (
979        X_extr : Xnew,
980        check : false
```

```
981        ) else (
982          if (Xnew <= Xmax and Xnew>=Xmin) then (
983            Xold : Xnew
984          ) else (
985            if (func(Xmin)<=func(Xmax)) then (
986              X_extr : Xmin
987            ) else (
988              X_extr : Xmax
989            ),
990            check : false
991          )
992        )
993      ) ),
994    return(X_extr)
995  )$
996
997  /*
998  Constrained functions of one variable
999  The exterior penalty function method
1000
1001  Inputs:
1002  Xmin : Minimum value of X
1003  Xmax : Maximum value of X
1004  X0 : Start value
1005  n : Total number of function evaluations
1006  alpha : Scaling parameter
1007  r_p_0 : Initial penalty factor
1008  gamma_value : Scaling parameter for r_p
1009
1010  Output:
1011  X_extr_new : The extremum value
1012  */
1013  one_variable_constrained_exterior_penalty(Xmin, Xmax, X0, n, alpha,
         r_p_0, gamma_value) :=
1014  block([r_p],
1015    /* tolerance */
1016    tol : 0.001,
1017
1018    /* if g and h exist, calculate p(x) and q(x) */
1019    if (unknown(g[1])) then (
1020      len_g : length(listarray(g)),
1021      /*p(x) := sum(unit_step(g[m](x))*g[m](x)^2,m,1,len_g)*/
1022      p(x) := sum(max(0,g[m](x))^2,m,1,len_g)
1023    ),
1024    if (unknown(h[1])) then (
1025      len_h : length(listarray(h)),
1026      q(x) := sum(h[k](x)^2,k,1,len_g)
1027    ),
1028
1029    /* determine the pseudo-objective function */
1030    func(x) := f(x),
1031    check : true,
1032    X_extr_old : X0,
1033    r_p : r_p_0,
```

```
1034   while (check) do (
1035     if ( unknown(g[1]) and unknown(h[1]) ) then (
1036       kill(func),
1037       func(x) := f(x) + r_p*p(x) + r_p*q(x)
1038     ) else if ( unknown(g[1]) and not(unknown(h[1])) ) then (
1039       kill(func),
1040       func(x) := f(x) + r_p*p(x)
1041     ) else if ( not(unknown(g[1])) and unknown(h[1]) ) then (
1042       kill(func),
1043       func(x) := f(x) + r_p*q(x)
1044     ),
1045     [X_extr_new,n_iter] : bf_ver2_varN(Xmin, Xmax, X0, n, alpha,
         false),
1046     printf(true, "~% r_p: ~,6f , X_extr: ~,6f , Number of iterations:
         ~d", r_p, X_extr_new, k),
1047     if ( abs(X_extr_new - X_extr_old) < tol or is(gamma_value=1) )
         then (
1048       check : false
1049     ),
1050     X_extr_old : X_extr_new,
1051     r_p : r_p * gamma_value
1052   ),
1053   return(X_extr_new)
1054 )$
1055
1056 /*
1057 Constrained functions of one variable
1058 The interior penalty function method
1059
1060 Inputs:
1061 Xmin : Minimum value of X
1062 Xmax : Maximum value of X
1063 X0 : Start value
1064 n : Total number of function evaluations
1065 alpha : Scaling parameter
1066 r_p_0_prime : Initial penalty factor for inequality constaints
1067 r_p_0 : Initial penalty factor for equality constraints
1068 gamma_value : Scaling parameter for r_p
1069 cal_type : Calculation type (Fractional or Logarithmic)
1070
1071 Output:
1072 X_extr_new : The extremum value
1073 */
1074 one_variable_constrained_interior_penalty(Xmin, Xmax, X0, n, alpha,
      r_p_0, r_p_0_prime, gamma_value, cal_type) :=
1075 block([r_p],
1076   /* tolerance */
1077   tol : 0.001,
1078
1079   /* if g and h exist, calculate p(x) and q(x) */
1080   if (unknown(g[1])) then (
1081     len_g : length(listarray(g)),
1082     if (equal(cal_type,"Fractional")) then (
1083       p(x) := sum(-1/(g[m](x)),m,1,len_g)
```

```
1084        ) else if (equal(cal_type,"Logarithmic")) then (
1085          p(x) := sum(-log(-g[m](x)),m,1,len_g)
1086        )
1087     ),
1088     if (unknown(h[1])) then (
1089       len_h : length(listarray(h)),
1090       q(x) := sum(h[k](x)^2,k,1,len_g)
1091     ),
1092     func(x) := f(x),
1093     check : true,
1094     X_extr_old : X0,
1095     r_p : r_p_0,
1096     r_p_prime : r_p_0_prime,
1097     while (check) do (
1098       if ( unknown(g[1]) and unknown(h[1]) ) then (
1099         func(x) := f(x) + r_p_prime*p(x) + r_p*q(x)
1100       ) else if ( unknown(g[1]) and not(unknown(h[1])) ) then (
1101         func(x) := f(x) + r_p_prime*p(x)
1102       ) else if ( not(unknown(g[1])) and unknown(h[1]) ) then (
1103         func(x) := f(x) + r_p*q(x)
1104       ),
1105       [X_extr_new,n_iter] : bf_ver2_varN(Xmin, Xmax, X0, n, alpha,
              false),
1106       printf(true, "~% X_0: ~,6f , r_p_prime: ~,6f , X_extr: ~,6f ,
              Number of iterations: ~d", X0, r_p_prime, X_extr_new, k),
1107       if ( abs(X_extr_new - X_extr_old) < tol or is(gamma_value=1) )
              then (
1108         check : false
1109       ),
1110       X_extr_old : X_extr_new,
1111       r_p : r_p * gamma_value
1112     ),
1113     return(X_extr_new)
1114   )$
1115
1116   /*
1117   Sign determination for the range detection algorithm
1118   Uses the mid-point of each range to find the sign (positive, negative)
            of the function in that range
1119   The main idea here is that the function will not change its sign in
            the ranges defined by its roots
1120
1121   Inputs:
1122   X1 : Minimum X value in the range
1123   X2 : Maximum X value in the range
1124
1125   Output:
1126
1127   */
1128   one_variable_constrained_one_variable_constrained_sign_detector(X1,X2)
            :=
1129   block( [],
1130       /* probe point */
1131       xprob : (X1+X2)/2,
```

```
/* sign detection */
kill(func,p,q,temp_g,temp_h),
zerofun(X) := 0,
temp_g : copy(g),
if (unknown(temp_g[1])) then (
  /* remove not-participating constrains from g */
  len_g : length(listarray(temp_g)),
  for k : 1 thru len_g do (
    temp_g[k] : copy(g[k]),
    if not(sign(temp_g[k](xprob)) = pos) then (
      temp_g[k] : copy(zerofun)
    )
  ),
  len_g : length(listarray(temp_g)),
  /*p(X) := sum(unit_step(g[m](X))*g[m](X)^2,m,1,len_g)*/
  p(X) := sum((temp_g[m](X))^2,m,1,len_g)
),
temp_h : copy(h),
if (unknown(h[1])) then (
  /* remove not-participating constrains from h */
  len_h : length(listarray(temp_h)),
  for k : 1 thru len_h do (
    temp_h[k] : copy(h[k]),
    if not(sign(temp_h[k](xprob)) = pos) then (
      temp_h[k] : copy(zerofun)
    )
  ),
  len_h : length(listarray(temp_h)),
  q(X) := sum(h[k](X)^2,k,1,len_h)
),
func(X) := f(X),
if ( unknown(g[1]) and unknown(h[1]) ) then (
  kill(func),
  func(X) := f(X) + r_p*p(X) + r_p*q(X)
) else if ( unknown(g[1]) and not(unknown(h[1])) ) then (
  kill(func),
  func(X) := f(X) + r_p*p(X)
) else if ( not(unknown(g[1])) and unknown(h[1]) ) then (
  kill(func),
  func(X) := f(X) + r_p*q(X)
),
return()
)$

/*
Range detection algorithm for constrained problems with one variable
To show the function in different ranges
These ranges are defined based on the roots of the objective function

needs load(to_poly_solve),  /* to check if all the roots are real
     (isreal_p(x))) */

Inputs:
```

```
1185   Output:
1186   Prints out the available ranges
1187   */
1188   one_variable_constrained_range_detection() :=
1189   block([],
1190     root_vasymps_s : makelist(),
1191     if (unknown(g[1])) then (
1192       len_g : length(listarray(g)),
1193       for i : 1 thru len_g do (
1194         /* find the roots */
1195         roots_g : solve(num(ratsimp(g[i](X)))=0,X),
1196         /* find the vertical asymptotes */
1197         vasymps_g : solve(denom(ratsimp(g[i](X)))=0,X),
1198         for j : 1 thru length(roots_g) do (
1199           if (isreal_p(roots_g[j]) and rhs(roots_g[j])>Xmin and
1200   rhs(roots_g[j])<Xmax) then (
1201             root_vasymps_s : cons(rhs(roots_g[j]),root_vasymps_s)
1202           )
1203         ),
1204         for j : 1 thru length(vasymps_g) do (
1205           if (isreal_p(vasymps_g[j]) and rhs(vasymps_g[j])>Xmin and
1206   rhs(vasymps_g[j])<Xmax) then (
1207             root_vasymps_s : cons(rhs(vasymps_g[j]),root_vasymps_s)
1208           )
1209         )
1210       )
1211     ),
1212     if (unknown(h[1])) then (
1213       len_h : length(listarray(h)),
1214       for i : 1 thru len_h do (
1215         /* find the roots */
1216         roots_h : solve(num(ratsimp(h[i](X)))=0,X),
1217         /* find the vertical asymptotes */
1218         vasymps_h : solve(denom(ratsimp(h[i](X)))=0,X),
1219         for j : 1 thru length(roots_h) do (
1220           if (isreal_p(roots_h[j]) and rhs(roots_h[j])>Xmin and
1221   rhs(roots_h[j])<Xmax) then (
1222             root_vasymps_s : cons(rhs(roots_h[j]),root_vasymps_s)
1223           )
1224         ),
1225         for j : 1 thru length(vasymps_h) do (
1226           if (isreal_p(vasymps_h[j]) and rhs(vasymps_h[j])>Xmin and
1227   rhs(vasymps_h[j])<Xmax) then (
1228             root_vasymps_s : cons(rhs(vasymps_h[j]),root_vasymps_s)
1229           )
1230         )
1231       )
1232     ),
1233     kill(r_p),
1234     /* remove duplicate roots and asymptotes */
       root_vasymps_temp : setify(root_vasymps_s),
       /* assign root_vasymps_temp arguments to root_vasymps_s */
       root_vasymps_s : args(root_vasymps_temp),
       /* sort the components from the lowest to the highest */
```

```
1235      root_vasymps_s : sort(root_vasymps_s,ordermagnitudep),
1236      print(" "),
1237      printf(true, "~% For ~,6f < X < ~,6f :", Xmin, root_vasymps_s[1]),
1238      print(" "),
1239      /* Calcualtion of the pseudo-function for different ranges */
1240      one_variable_constrained_one_variable_constrained_sign_detector(Xmin,
1241      root_vasymps_s[1]),
1242      print(%Phi, "=", func(X)),
1243      len_root : length(root_vasymps_s),
1244      if (len_root > 1) then (
1245        for i : 1 thru (len_root-1) do(
1246          if not(root_vasymps_s[i]=root_vasymps_s[i+1]) then (
1247            print(" "),
1248            printf(true, "~% For ~,6f < X < ~,6f :", root_vasymps_s[i],
1249          root_vasymps_s[i+1]),
              print(" "),
1250            one_variable_constrained_one_variable_constrained_sign_detector
1251            (root_vasymps_s[i],
1252            root_vasymps_s[i+1]),
1253            print(%Phi, "=",func(X))
1254          )
1255        )
1256      ),
1257      print(" "),
1258      printf(true, "~% For ~,6f < X < ~,6f :", root_vasymps_s[len_root],
          Xmax),
1259      print(" "),
1260      one_variable_constrained_one_variable_constrained_sign_detector
1261      (root_vasymps_s
1262      [len_root],Xmax),
1263      print(%Phi, "=",func(X)),
1264      kill(root_vasymps_s),
1265      return()
1266    )$
1267
1268    /*
1269    Pseudo-function calculator for interior penalty method
1270
1271    Inputs:
1272    cal_type : type of calculation for different formulations
1273    Fractional or Logarithmic
1274
1275    Output:
1276    */
1277    pseudo_function_interior_penalty(cal_type) :=
1278    block([],
1279
1280      /* if g and h exist, calculate p(x) and q(x) */
1281      if (unknown(g[1])) then (
1282        len_g : length(listarray(g)),
1283        if (equal(cal_type,"Fractional")) then (
1284          p(x) := sum(-1/(g[m](x)),m,1,len_g)
1285        ) else if (equal(cal_type,"Logarithmic")) then (
1286          p(x) := sum(-log(-g[m](x)),m,1,len_g)
```

```
1287        )
1288      ),
1289
1290    if (unknown(h[1])) then (
1291      len_h : length(listarray(h)),
1292      q(x) := sum(h[k](x)^2,k,1,len_g)
1293    ),
1294    func(x) := f(x),
1295    if ( unknown(g[1]) and unknown(h[1]) ) then (
1296      func(x) := f(x) + r_p_prime*p(x) + r_p*q(x)
1297    ) else if ( unknown(g[1]) and not(unknown(h[1])) ) then (
1298      func(x) := f(x) + r_p_prime*p(x)
1299    ) else if ( not(unknown(g[1])) and unknown(h[1]) ) then (
1300      func(x) := f(x) + r_p*q(x)
1301    ),
1302    return()
1303  )$
1304
1305  /*
1306  Constrained functions of several variables
1307  Using the Steepest Descent Method
1308  The exterior penalty function method
1309
1310  Inputs:
1311  func_obj : Objective function
1312  no_of_vars : Total number of variables
1313  X_0 : Vector of start values
1314  alpha_0 : Initial scaling parameter (alpha)
1315  gamma_value : Value of the scaling parameter (gamma)
1316  eps : Tolerance value
1317  criterion : Convergence criterion  (max_iter, abs_change, rel_change
             or Kuhn_Tucker)
1318  print_switch : Prints the output if true
1319
1320  Output:
1321  x_extr_new : The extremum value
1322  f_extr_new : The function value at the extremum
1323  */
1324  steepest_multi_variable_constrained(func_obj,no_of_vars,X_0,
1325  alpha_0,eps,r_p_0, gamma_value,criterion,print_switch) :=
1326  block([ i, r_p, X_new_value, X_old_value],
1327
1328    for i:1 thru no_of_vars do X[i] : X[i],
1329
1330    /* if g and h exist, calculate p(x) and q(x) */
1331    if (unknown(g[1])) then (
1332      len_g : length(listarray(g)),
1333      p(X) := sum(unit_step(g[m](X))*g[m](X)^2,m,1,len_g),
1334      p_grad(X) := sum(unit_step(g[m](X))*gradient(g[m](X)^2,
             listarray(X)),m,1,len_g)
1335    ),
1336
1337    if (unknown(h[1])) then (
1338      len_h : length(listarray(h)),
```

```
q(X)  := sum(h[k](X)^2,k,1,len_h),
q_grad(X) := sum(gradient(h[k](X)^2,listarray(X)),k,1,len_h)
),

/* determine the pseudo-objective function */
func(X) := func_obj(X),
func_grad : gradient(func(X), listarray(X)),

check_constrained : true, /* convergence check variable for the
  constrained problem */
x_extr_old : copy(X_0),
r_p : r_p_0,
while (check_constrained) do (
  if ( unknown(g[1]) and unknown(h[1]) ) then (
    kill(func,func_grad),
    func(X) := func_obj(X) + r_p*p(X) + r_p*q(X),
    func_grad : gradient(func(X), listarray(X)) + r_p*p_grad(X) +
    r_p*q_grad(X)
  ) else if ( unknown(g[1]) and not(unknown(h[1])) ) then (
    remvalue(func,func_grad),
    func(X) := func_obj(X) + r_p*p(X),
    func_grad : gradient(func_obj(X), listarray(X)) + r_p*p_grad(X)
  ) else if ( not(unknown(g[1])) and unknown(h[1]) ) then (
    kill(func,func_grad),
    func(X) := func_obj(X) + r_p*q(X),
    func_grad : gradient(func(X), listarray(X)) + r_p*q_grad(X)
  ),

/* send the problem as an unconstrained one to the corresponding
  solving routine */
  [counter,x_extr_new,f_extr_new] :
  steepest_multi_variable_unconstrained_var2(func_obj,func,func_grad,
  no_of_vars,X_0,alpha_0,eps,criterion,print_switch),

/* check for convergence of the solution */
  kill(X_old_value,X_new_value),
  for i:1 thru no_of_vars do X_new_value[i] : X[i] =
  rhs(x_extr_new[i]),
  for i:1 thru no_of_vars do X_old_value[i] : X[i] =
  rhs(x_extr_old[i]),
  X_new_value : listarray(X_new_value),
  X_old_value : listarray(X_old_value),
  f_new_constrained : at(func,X_new_value),
  f_old_constrained : at(func,X_old_value),
  if ( abs(f_new_constrained - f_old_constrained) < eps or
  is(gamma_value=1) ) then (
    check_constrained : false
  ),

  x_extr_old : copy(x_extr_new),

  r_p : r_p * gamma_value
),
```

```
1386    for i:1 thru no_of_vars do x_extr_new[i] : X[i] = rhs(x_extr_new[i]),
1387
1388    return([x_extr_new,f_extr_new])
1389  )$
1390
1391  /*
1392  Unconstrained functions of several variables
1393  Using the Steepest Descent Method
1394
1395  Inputs:
1396  func : Objective function
1397  no_of_vars : Total number of variables
1398  X_0 : Vector of start values
1399  eps : Tolerance value
1400  criterion : Convergence criterion  (max_iter, abs_change, rel_change
          or Kuhn_Tucker)
1401  print_switch : Prints the output if true
1402
1403  Output:
1404  X_new : The extremum value
1405  */
1406  steepest_multi_variable_unconstrained(func,no_of_vars,X_0,alpha_0,eps,
1407  criterion,print_switch) :=
1408  block([i, X_new],
1409
1410    for i:1 thru no_of_vars do X[i] : X[i],
1411
1412    /* calculate the gradient of the function (included in this
          libraray) */
1413    f_grad : gradient(func(X), listarray(X)),
1414
1415    X_old : copy(X_0),
1416
1417    check : true,
1418    once : true,
1419    counter : 0,
1420    while (check) do (
1421      counter : counter + 1,
1422
1423    /* clean memory from old values */
1424      kill(X_old_value,X_new_value,X_new),
1425
1426    /* assign X_old_value */
1427      for i:1 thru no_of_vars do X_old_value[i] : rhs(X_old[i]),
1428      X_old_value : transpose(listarray(X_old_value)),
1429
1430    /* calculate the gradient and S_old*/
1431      f_grad_value : float(at(f_grad, X_old)),
1432      S_old : -1*f_grad_value,
1433      S_old : transpose(S_old),
1434
1435    /* find alpha_star */
1436      alpha_star : alpha_old_finder(func, X_old_value, S_old, alpha_0,
          eps, counter),
```

```
      /* calculate X_new_value */
      for i:1 thru no_of_vars do X_new[i] : X_old[i] +
      alpha_star*S_old[i][1],
      X_new : listarray(X_new),
      for i:1 thru no_of_vars do X_new_value[i] : X[i] = rhs(X_new[i]),
      X_new_value : listarray(X_new_value),

      /* check if the solution is converged */
      f_old : at(func(X),X_old),
      f_new : at(func(X),X_new_value),
      if (print_switch) then (
        print(i=counter,X=X_new_value,"func(X)"=f_new)
      ),
      /* convergence criterion */
      if (criterion = "max_iter") then (
        if (counter=maximum_iteration) then check : false
      ) elseif (criterion = "abs_change") then (
        if (abs(f_new-f_old)<eps) then check : false
      ) elseif (criterion = "rel_change") then (
        if ((abs(f_new-f_old)/max(f_new,10^-10))<eps) then check : false
      ) elseif (criterion = "Kuhn_Tucker") then (
        f_grad_value_new : float(at(f_grad, X_new_value)),
        check : false,
        for i:1 thru no_of_vars do (
          if (abs(f_grad_value_new[i])>eps) then check : true
        )
      ),

      /* update X_old if the solution is not converged*/
      if (check) then (
        for i:1 thru no_of_vars do X_old[i] : X[i] = rhs(X_new[i]),
        alpha_0 : copy(alpha_star)
      )
    ),
    printf(true,"Converged after ~d iterations!", counter),
    return(X_new)
  )$

/*
Constrained functions of several variables
Using the Steepest Descent Method
Variation 2
The exterior penalty function method

Purpose:
This function works the same as
      *steepest_multi_variable_unconstrained*, with the difference that
      now the pseudo-objective function is also an input

Inputs:
func_obj : Objective function
func : Pseudo-objective function
func_grad : Gradient of the pseudo-objective function
```

```
1488  no_of_vars : Total number of variables
1489  X_0 : Vector of start values
1490  alpha_0 : Initial scaling parameter (alpha)
1491  eps : Tolerance value
1492  criterion : Convergence criterion  (max_iter, abs_change, rel_change
          or Kuhn_Tucker)
1493  print_switch : Prints the output if true
1494
1495  Output:
1496  counter : Total number of iterations required for convergence
1497  X_new : The extremum value
1498  f_new : The function value at the extremum
1499  */
1500  steepest_multi_variable_unconstrained_var2(func_obj,func,func_grad
1501  ,no_of_vars,X_0,alpha_0,eps,criterion,print_switch) :=
1502  block([i, X_new],
1503
1504    for i:1 thru no_of_vars do X[i] : X[i],
1505
1506    X_old : copy(X_0),
1507
1508    alpha_0_store : copy(alpha_0),
1509
1510    check : true,
1511    once : true,
1512    counter : 0,
1513
1514    while (check) do (
1515      counter : counter + 1,
1516
1517    /* clean memory from old values */
1518      kill(X_old_value,X_new_value,X_new),
1519
1520    /* assign X_old_value */
1521      for i:1 thru no_of_vars do X_old_value[i] : rhs(X_old[i]),
1522      X_old_value : transpose(listarray(X_old_value)),
1523
1524    /* calculate the gradient and S_old*/
1525      f_grad_value : float(at(func_grad, X_old)),
1526      S_old : -1*f_grad_value,
1527      S_old : transpose(S_old),
1528
1529    /* find alpha_star */
1530      alpha_star : alpha_old_finder_var2(func_obj, X_old_value, S_old,
          alpha_0_store, eps, counter),
1531
1532    /* calculate X_new_value */
1533      for i:1 thru no_of_vars do X_new[i] : X_old[i] +
          alpha_star*S_old[i][1],
1534      X_new : listarray(X_new),
1535      for i:1 thru no_of_vars do X_new_value[i] : X[i] = rhs(X_new[i]),
1536      X_new_value : listarray(X_new_value),
1537
1538    /* check if the solution is converged */
```

```
1539      f_old : at(func(X),X_old),
1540      f_new : at(func(X),X_new_value),
1541      if (print_switch) then (
1542        print(i=counter,X=X_new_value,"func(X)"=float(f_new))
1543      ),
1544      /* convergence criterion */
1545      if (criterion = "max_iter") then (
1546        if (counter=maximum_iteration) then check : false
1547      ) elseif (criterion = "abs_change") then (
1548        if (abs(f_new-f_old)<eps) then check : false
1549      ) elseif (criterion = "rel_change") then (
1550        if ((abs(f_new-f_old)/max(f_new,10^-10))<eps) then check : false
1551      ) elseif (criterion = "Kuhn_Tucker") then (
1552        f_grad_value_new : float(at(func_grad, X_new_value)),
1553        check : false,
1554        for i:1 thru no_of_vars do (
1555          if (abs(f_grad_value_new[i])>eps) then check : true
1556        )
1557      ),
1558
1559      /* update X_old if the solution is not converged*/
1560      if (check) then (
1561        for i:1 thru no_of_vars do X_old[i] : X[i] = rhs(X_new[i]),
1562        alpha_0_store : copy(alpha_star)
1563      )
1564    ),
1565    printf(true,"Converged after ~d iterations!", counter),
1566    return([counter,X_new,f_new])
1567  )$
```

Index

Printed in the United States
by Baker & Taylor Publisher Services